EVEREST

Mountain Without Mercy

EVEREST

Mountain Without Mercy

BROUGHTON COBURN

INTRODUCTION BY TIM CAHILL
AFTERWORD BY DAVID BREASHEARS

NATIONAL
GEOGRAPHIC
SOCIETY

For the Venerable Ngawang Tenging Zangbu,
The Incarnate Lama of the Tengboche Monastery

Major funding for the MacGillivray Freeman Film *Everest* was provided by the
National Science Foundation*

and the
Everest Film Network

Science World British Columbia   Museum of Science, Boston
Fort Worth Museum of Science and History   National Museum of Natural Science, Taichung
Houston Museum of Natural Science   Denver Museum of Natural History

* The opinions expressed herein are those of the authors and not necessarily those of the Foundation.

Library of Congress Cataloging-in-Publication Data

Coburn, Broughton, 1951-
  Everest : mountain without mercy / by Broughton Coburn;
introduction by Tim Cahill, afterword by David Breashears.
       p.   cm.
  Includes index.
  ISBN 0-7922-7014-2
  1. Mount Everest Expedition (1996) 2. Mount Everest Expedition
(1996)—Pictorial works. 3. Mountaineering accidents—Everest,
Mount (China and Nepal) 4. Mountaineering accidents—Everest, Mount
(China and Nepal)—Pictorial works. 5. IMAX Corporation.
I. Title.
GV199.44.E85C63 1997
796.5'22'095496—dc21   97-10765
                                                    CIP

PRECEDING PAGES:

PAGES 2-3	PAGES 4-5	PAGES 6-7	PAGES 8-9
Seeming as unattainable as the mythical Olympus, a cloudless Everest rises above and beyond the Nuptse-Lhotse ridge, towering over trekkers on a foothill trail along the traditional southern approach to the Khumbu region and to Everest.	*Icy highway to Everest's lower reaches: Stark sweeps of the Western Cwm—a broad, glacier-filled extension of the Khumbu valley—dwarf climbers bound for Camp II.*	*Aglow from within, climbers' lantern-illuminated tents nestle at Base Camp, more than a vertical mile below the moonlit shoulders of Everest (left) and Nuptse.*	*Often capped by cloud, Everest makes its own weather while it lures adventurers with the ultimate challenge. To date, more than 4,400 people have sought its top; only 728 have reached the summit.*

TABLE OF CONTENTS

SORROW ON TOP OF THE WORLD

BY TIM CAHILL

I was sitting at my desk, working on a documentary screenplay about Mount Everest, trying to deal in the rarefied realms of the scientific and the sacred, when the phone rang and my fax machine buzzed its electronic mating call. The message was from Broughton Coburn, a friend and colleague living in Wilson, Wyoming. His information, I learned later, was based on a report by another colleague, Audrey Salkeld, and had come direct, via satellite, from Nepal. Terrible and tragic news inched its way up the paper, line by line:

FROM: Broughton Coburn
Dispatch 2 : May 11
Tragedy
There has been a disaster on Mt. Everest.
Breashears is…

Messengers from on high? Soaring ravens—believed by Sherpas to be bearers of omens—converge over prayer flags adorning a traditional shrine at Everest Base Camp.

And I waited for what seemed to be a very long time for the next line to come up. David Breashears was the co-director/producer and cinematographer of *Everest,* the large-format documentary film I was working on about an international team of climbers attempting to install the world's highest Global Positioning System and weather station on the South Col of Mount Everest. If conditions warranted, they would try to summit and bring back the first IMAX® camera footage from the top of the world. No one had ever attempted this before because everyone knew it was impossible.

Everyone, that is, except David Breashears, who was, in fact, currently shooting the climbing sequences somewhere on the mountain. I had met with him a few months earlier. The man was a tireless worker, entirely professional. He was also entirely too intense for some people's taste. In the several days we'd spent working out concepts on the script, I had come to respect and admire David Breashears absolutely. We worked well together. I don't want to overdramatize this: The thing I felt as the fax paper crawled up my machine was just a flash —bang, like that—the kind of sudden lurching you feel in your gut when a child runs out in front of your car. A quick turn of the wheel and everything's OK— disaster averted—but minutes later your hands still shake and your heart pounds and your breath comes harshly in your throat.

Breashears, the fax continued,

is safe but is joining the rescue effort.

The entire team—climbers and filmmakers—it seemed, was safe. But eight others had died in a storm. And another had fallen to his death a day earlier.

Over several harrowing days in May 1996, the entire world watched as terror and triumph played themselves out like a classical Greek tragedy on the wind-whipped, snow-scoured stage of the world's highest mountain.

I had been working on the *Everest* screenplay for several months, and had discovered that writing a large-format documentary film is strange work. It's not like a video shoot, where most everything can be filmed at any time or at a moment's notice. Because the IMAX camera is so heavy, the setups so complex, and the film itself so expensive, there are an extremely limited number of shots planned and, for this reason, the shooting script must be prepared well ahead of time. That way, the director knows exactly where the camera goes and precisely what will be important. The before-the-fact screenplay is supposed to cut down on unnecessary expense. In practice, the filmmakers go to work, and, of course, nothing happens the way it was planned and certainly not the way it was written. Other more intense shots than those envisioned occur, events rumble over speculation like an avalanche, truth proves more compelling than guesswork. So when the crew returns from locations with scenes and stories that have nothing whatever to do with the original script, the entire screenplay is rewritten, top to bottom, as it should be. The shooting script is one person's fantasy. The raw footage shot on location is a documentary effort that is then massaged and stroked— sometimes pummeled and hammered—into a final product that can be, in the best cases, evocative, challenging, beautiful, exhilarating, humorous, and even emotional. My plan for the shooting script had been to combine science with adventure, and even humor. The humor was meant to dissolve tension in what was, after all, a life-or-death undertaking. This humor, which even I had to admit wasn't very funny, mostly involved Spam.

The joke? Well, one of our expedition's climbers, "Steady Ed" Viesturs, a man universally thought to be among the strongest high-altitude mountaineers on earth, liked Spam. He'd climbed Mount Everest before, three times actually, twice without the use of supplemental oxygen, but never without the supplemental use of Spam. He called it high-altitude climbing fuel. That, you see, was the joke. Actually, I'd introduced the hilarity of Spam two drafts into the shooting script. Steve Judson and I were writing the screenplay for MacGillivray Freeman Films, longtime producers of large-format motion pictures. The film itself was partially funded by the National Science Foundation and would be shown primarily in museum IMAX theaters. The NSF required a strong—and highly accurate—science content. To that end, a phalanx of science advisers—geologists, physiologists, ethnologists—pored over every draft, firing off a whiteout blizzard of corrections, suggestions, and sometimes exasperated comments.

I recall experiencing a certain amount of despair as I pored over my notes on the script from various advisers. Not surprisingly, the geologists wanted more geology; the physiologist wanted more physiology; the ethnologist wanted more ethnology. Brad Washburn, the respected director emeritus of the Boston Museum of Science, thought that the film was

too heavy on science, and that climbing Mount Everest was the last great adventure on earth. As filmmakers, we should concentrate on the adventure and let the science work its way into the film when appropriate.

This was, of course, entirely contradictory advice. Unfortunately, I agreed with everyone, including one of the distinguished geologists on the project who had marked up an early draft of the script with his comments on my geological solecisms. Stepping out of his own field of expertise for a moment, he had written, "Spam...oh please."

The guy was probably right. At the time, the climbing team was already at Everest Base Camp, acclimatizing and waiting for a break in the weather before beginning their ascent. Did they really need to waste their time on a Spam scene? I went through the script and brutally killed every reference to canned meat. This Spam-bereft draft of the screenplay—the shooting script—went directly to the Everest Base Camp, via satellite e-mail. Not, I thought, that it would have much effect on the climbers and filmmakers there. They were going to do what they had to do to survive, accomplish their scientific mission, and possibly struggle to the summit of the mountain. The climbing team consisted of two women—Araceli Segarra of Spain and Sumiyo Tsuzuki of Japan—along with two men—Ed Viesturs of the United States and Jamling Norgay of India, a Sherpa and son of Tenzing Norgay, the first man (along with Sir Edmund Hillary) to stand on the summit of Everest.

A month after the final shooting script arrived at Base Camp, news of the tragedy on Everest came beeping through my fax machine.

Our team, it became clear very early, was safe. It took me almost 12 more hours to learn that Jon Krakauer (a friend and colleague at *Outside* magazine) had survived, and even reached the summit, on the very day that several other climbers perished in a storm atop the world's highest mountain.

The behavior of the film crew, climbers, and filmmakers, was exemplary. Breashears, his assistant cameraman, Robert Schauer, an Austrian who had climbed Everest once before, and Ed Viesturs recovered one body, and helped save the lives of two injured climbers. Aside from the cumbersome IMAX camera gear, the film team also carried video cameras. It was apparent that this tragedy was of interest worldwide. The filmmakers chose to save lives rather than shoot what would likely be very lucrative video. So professionally driven David Breashears proved himself a hero, not only physically but also morally. Meanwhile, Sumiyo Tsuzuki helped tend to injured climbers at Advanced Base Camp, and Araceli Segarra came up with an idea that facilitated one of the world's highest helicopter medical evacuations.

At this point, the team had to descend or die. Above 21,000 feet, the body simply does not acclimatize to altitude. Emotionally and physically exhausted, the team returned to Base Camp, at the foot of the Khumbu Glacier, where they attended a memorial service for the perished climbers. Meanwhile, the monsoon season was approaching, and with it, the end of any hope of climbing the mountain that year.

Decisions had to be made. Greg MacGillivray, president of MacGillivray Freeman Films, and co-director/producer of the film, was in touch with the team via satellite phone. His instructions were simple and to the point: Nothing—no amount of funding or professional pride—was worth one more death on the mountain. In retrospect, however, it seems Breashears never even considered retreat: If conditions improved, he'd complete the scientific mission and try for the summit. Both the film team and the climbing team were spectacularly successful. Jamling Norgay honored his father's memory on the summit of Everest while Araceli Segarra became the first Spanish woman to climb the mountain. Ed Viesturs summited a fourth time, once again without the use of supplemental oxygen. Breashears, for his part, made cinematic history: He brought back the first high-definition, large-format footage from the summit of Mount Everest, shooting two complete rolls of film, risking frostbite, and worse, by changing the rolls barehanded.

Meanwhile, much was being made of the deaths on Mount Everest. It was the lead item on nightly network news, and newsmagazine pundits were already reflecting on the catastrophe. A few suggested that the tragedy was simply a combination of bad luck, bad weather, and poor decision making. Mostly, fingers were pointed as various climbing pundits weighed in with their opinions. Blame—there had to be someone to blame, didn't there?—was dished out, appropriately or not, depending on one's view of the events.

The problem of motivation was often invoked. Guides, some said, were greedy and accepted climbing clients who had no business attempting Everest. Certain climbers—those paying up to $65,000 apiece for the chance to risk their lives on the mountain—were characterized as glory hounds. The mountain, some Sherpas said, was angry. Many Western and Japanese critics agreed.

Who, the question was asked, actually "belongs" on the mountain? Were skill and stamina and will enough? Or was the summit like some children's storybook paradise where only the pure of heart and the well intentioned were admitted? Was there a spiritual component to climbing? Vexing questions.

As editors at MacGillivray Freeman Films worked on Breashears' historic film footage, it became obvious to everyone that the complete story could not be told in a 40-minute screenplay. A book could more completely cover the science, the controversy, the triumph and the tragedy as well as the various events reluctantly left on the cutting room floor.

A writer was needed and Broughton Coburn was the obvious choice. Brot has lived in Nepal off and on for over 17 years. He had worked with Breashears on the *Everest* film, handling logistics and translation. A climber himself, Brot first went to Nepal in 1973 with the Peace Corps, after graduating from Harvard. As a practicing Buddhist and a student of Sherpa culture, he consulted with me on the screenplay, speaking often on the phone about various subjects, including "proper motivation." For what it's worth, Brot—the only known human being to suffer the attentions of a Nepalese nose leech—did not object to most of the Spam in my original version of the screenplay, which indicates that he is a man of some humor, though probably not a gourmet. More to the point, I can say that Brot, along with his friend Jamling Norgay, helped all of us who worked on the film to comprehend more fully the Sherpa people and the spiritual aspects of climbing. In the end, that is what this story is about: strands of heroism and tragedy, science and spirit, all braided together and wound about various troubling questions of motivation and purpose. Brot Coburn is uniquely qualified to deal with such complex issues and attempt to answer the most vexing of these questions.

PROLOGUE

On the morning of May 10, the steep, knife-edge ridge connecting the South Summit of Mount Everest to the Hillary Step was clearly visible from Camp II, 7,300 vertical feet below. High clouds raced past the summit, and high winds scoured the Southeast Ridge, emitting a distant, ominous roar. ¶ In the early afternoon, David Breashears and Ed Viesturs trained their binoculars on the South Summit and spotted six or seven climbers traversing the ridge. Directly ahead, several more climbers were waiting at the base of the Hillary Step, an awkward and craggy stretch of 40 vertical feet of rock and snow leading to the summit that can only be managed one climber at a time. David and Ed would later learn that the ropes needed to assist climbers had been fixed on the Step later in the morning than guides Rob Hall and Scott Fischer had planned. ¶

Traffic jam near the top: Backed up by delays in the rigging of fixed ropes, climbers bottleneck on May 10, 1996 just below the Hillary Step, 40 vertical feet of treacherous ice and rock that can only be scaled one person at a time.

Even when fixed promptly, a time-consuming bottleneck can occur when too many climbers arrive at the Step at the same time.

Squinting into a telescope, David and Ed calculated the climbers' rate of ascent and their use of bottled oxygen: Their oxygen might last through the summit attempt, and the weather was fair. Nevertheless, Ed's face worked with concern. It was very late in the day for climbers to be attempting to summit. Because it can take 18 hours to make the round-trip from Camp IV to the top of Everest under the best of conditions, almost all climbers try to summit before midday. The climbers Ed and David observed would almost certainly be descending in the fading light of late afternoon—a potentially dangerous situation.

"They've already been climbing 14 hours, and they *still* aren't on the summit. Why haven't they turned around?" Ed asked Breashears. There was no answer. Viesturs was glad of one thing: that he wasn't waiting in line with them.

Ed and Paula Viesturs had been on Everest with New Zealander Rob Hall's group the two previous years, and Paula was chatting with Helen Wilton, Hall's manager at the New Zealand Base Camp. They were eager to hear news from the summit, and at about 1:25 p.m. whoops and cheers issued from the dining tent of American Scott Fischer's Mountain Madness group—his guides and clients had just radioed from the top. Fischer himself had said he was having a slow day. He had not reached the summit yet,

but apparently chose to continue climbing slowly rather than help guide his clients down. They passed him as they descended.

Neither Rob Hall nor his clients had made it to the top yet. Around 2:30 Paula and Helen heard Hall's summit radio call. It was windy, he reported; those on his team who had summited were descending, except for client Doug Hansen. He could see Hansen approaching slowly, and had decided to wait for him at the top. Paula was surprised that Rob would summit with a client this late, contrary to his own rule that his team should descend no later than 2:00 p.m.

Paula was tempted to join in the cheering, but thought it premature. She worried about Hansen, and was well aware of how dangerous and treacherous Everest could be. The year before, Chantal Mauduit was climbing behind Ed Viesturs when she collapsed at the South Summit while attempting to be the first woman to summit without supplemental oxygen. In an extremely difficult rescue, the Sherpas, Ed, Rob Hall, and two other climbers had to drag and lower her down the summit ridge to Camp IV on the South Col. On that climb Ed had also encountered Hall's client Doug Hansen staggering below the South Summit. Hansen was not clipped in to the fixed rope, and Ed shouted at him to stop. Ed gave Hansen his only oxygen bottle, but Hansen was still having trouble descending even as Guy Cotter, Hansen's guide, assisted him back to the South Col.

"Keep me informed," Paula told Helen, as she returned to her Base Camp kitchen tent to make dinner.

With symmetrical Tawoche as backdrop, rock cairns and carved mani stones along the trail signify this overlook as a "power spot."

A MOUNTAIN OF A DREAM

*"We had seen a whole mountain range, little by little, the lesser to the greater until,
incredibly higher in the sky than imagination had ventured to dream,
the top of Everest itself appeared."*

—GEORGE MALLORY

"Several seasons of good weather have led people to think of Everest as benevolent," *Everest* Film Expedition leader David Breashears observed. "But in the mid-eighties—before many of the *guides* had been on Everest— there were three consecutive seasons when no one climbed the mountain because of the ferocious wind. Everest can be a place where people can't see or move, where tents are ripped apart, where all the high-tech gear in the world can't save you." ¶ Yet Everest casts a spell, and many remain undeterred. More than 150 people have died on the mountain—1 for about every 30 attempting to climb it, or 1 for every 5 who have reached the summit. Most of the dead are still on the mountain. ¶ Mount Everest hasn't always attracted such misfortune. As far as is known, no one considered

Trademark plume streams east from Everest's Southeast Ridge—a sure sign of steady, heavy winds on top. Taken from Lhotse, this view shows the team's route, from the South Col to the Southeast Ridge and the summit.

climbing Everest until long after the British commissioned the Great Trigonometrical Survey, a monumental effort to survey and measure all of India and the Himalaya. Undertaken in the mid-19th century, the survey identified the world's tallest mountain, assigned it a provisional height, and gave it a name: Mount Everest, in honor of India's previous Surveyor General, George Everest.

AN UNUSUAL CONCEPT

Filmmaker Greg MacGillivray, a pioneer of large-format (IMAX) cinematography, had long been fascinated by Everest exploration and mountaineering—the dramatic story of human achievement and heroism set against insurmountable odds. He carefully studied climbing films, which he felt were limited in their ability to capture the splendor of the mountains and the spectacle of climbing them. Someone had to be able to do it better, and IMAX was the medium to match the Himalaya. In IMAX theaters, audiences are surrounded by six channels of synchronized digital sound, and scenes are projected at 10 times the resolution of a 35-mm feature film on screens that are up to 8 stories tall and 100 feet wide.

"Just as the wide screen of Cinemascope was perfect for the landscapes of John Ford's westerns of the 1950s," MacGillivray said, "the more vertical ratio aspect of IMAX is perfect for Everest."

Temporarily barehanded at 21,400 feet, Everest expedition leader David Breashears (opposite, in cap, and below) reloads film at Camp II with help from filmmaker and team member Robert Schauer.

In June of 1994, MacGillivray contacted David Breashears to ask if he might like to co-produce and co-direct a film on Everest—and organize the expedition to the mountain.

Breashears was the obvious choice. A veteran of 18 Himalayan expeditions, he was the first American to summit Everest twice, and has made eight films on and around the mountain. He was intrigued by MacGillivray's proposal, but concluded that hauling a full-size IMAX camera to the summit would be impossible.

Always game for a challenge within the realm of safety, however, Breashears figured he'd try to convince MacGillivray to shoot the high-mountain footage in a more manageable format, such as 35 mm. "I'm interested," he responded confidently. Breashears had transmitted the first live television images from Everest's summit 11 years earlier, and knew that a feature movie camera would be only slightly larger than a video camera.

But MacGillivray was firm; the images would have to be shot in IMAX. Smaller formats, he argued, wouldn't work on the big screens of the museum theaters that typically show these large-format films. He could see the logistical problems that transporting the camera posed, but wouldn't lower his expectations. The production company would simply have to build a smaller camera, and then provide the resources and support for an expedition that could make it happen.

THE ELUSIVE HEIGHT OF EVEREST

BY ROGER BILHAM, PROFESSOR OF GEOLOGY, UNIVERSITY OF COLORADO, BOULDER

On a cloudless morning in 1847 near the Ganges River, an officer of the Survey of India sat atop a masonry tower focusing a heavy brass theodolite. A hundred miles away, a snow-covered mountain appeared upside down in the telescope's eyepiece.

Across from the entry for "Peak XV," the officer recorded a row of eight numbers—the exact angle between the horizon and the peak's snowy summit. He moved on to Peak XVI.

In the early 1800s, the Great Trigonometrical Survey of India, the largest surveying project in history, was launched by the British. Its objective was to map India from its southern shores to the northern mountains, accurate to within a few feet. The precision of the survey relied on a theodolite grid of linked triangles, invisibly crisscrossing the country from hilltop to hilltop. Where no hills existed, towers were laboriously constructed at ten-mile intervals to see above the jungle and the curvature of the earth. Where ox-drawn carts couldn't pass through swamps and forests, elephants transported the survey equipment—including the 1,200-pound Great Theodolite. Eventually, sightings of Peaks XV and XVI and other Himalayan summits were made from five additional stations.

When this extraordinary survey was completed, the data were processed by "computers"—mathematicians assigned by the Survey of India, equipped with logarithms, pencils, and reams of paper. Two independent teams of computers sat on either side of a long table, and they contributed sequentially to each calculation. At the far end of the room, the two answers were compared. Their calculations unveiled steadily higher summits until they obtained 29,002 feet, for Peak XV.

The now famous phrase, "Sir! I have discovered the highest mountain in the world," attributed to the chief computer as he relayed the news to Surveyor-General Waugh, captures the excitement of otherwise dull work spread over many years. It was considered appropriate that Peak XV should be named after the father of Indian geodesy, Sir George Everest.

Although the early observations were precise to thirty millionths of a degree, they also

yielded the first of many incorrect height estimates. Even now, Mount Everest's precise altitude continues to elude surveyors.

To calculate the height of a distant mountain, three corrections must be made. The first factors the curvature of the earth. Though simple to compute, the shape of the earth was not well known in 1852. Second, the apparent position of the peak must be adjusted for the refraction of light as it descends at a shallow angle from the frigid peaks to the denser hot air above the Ganges plain. Unless observed from a nearby mountain, atmospheric refraction causes the mountain to appear much too high.

Finally, a mountain's height must be measured from sea level, which in fact isn't always "level." If a canal were dug from the sea to the base of Mount Everest, the water in it would be drawn upward by the gravitational attraction of the Himalaya. The early surveyors knew this, having noticed that the leveling bubbles of their theodolites shifted increasingly southward—relative to the fixed stars—as they approached the Himalaya. This meant that the geoid—the theoretical extension of sea level beneath the mountains—bulged upward. Estimating the geoid was the most problematic correction; unable to make crucial measurements in Nepal, the sur-

veyors had to guess the weight, and thereby the gravitational attraction, of the Himalaya and the Tibet Plateau.

Periodically, new corrections were applied to the old data. Finally, in 1950, the Survey of India was permitted to enter Nepal. Between 1952 and '54 B. L. Gulatee, Director of the Geodetic and Research Branch, took readings within 30 miles of Everest, using a remarkable system of crossed quadrilaterals. He measured 29,028 feet, with an uncertainty of 10 feet.

Because of variations in snow depth on the summit, a more accurate estimate appeared impossible—except that mountaineers could now climb up there and measure it. Theoretically. In 1975, a Chinese team placed a tripod on the top that they determined was 29,029 feet above the Yellow Sea, with an uncertainty of one foot. A pole driven into the snow beneath the tripod stopped at three feet. The 1954 Indian and the 1975 Chinese measurements agreed. That would seem to be the end of the quest.

A decade later, however, navigational satellites fitted with GPS transmitters made their first pass over the Himalaya. Using laser prisms and GPS receivers, a group of Italian surveyors determined that the heights of Everest and K2 were slightly higher than Survey of India estimates.

Meanwhile, on the other side

Perched on a bare knob above Base Camp, geologist Roger Bilham deploys a GPS receiver that will enable scientists to determine the site's precise latitude and longitude, and help them monitor positional changes wrought by tectonic forces.

of the world, veteran cartographer Bradford Washburn, supported by a research fund of Boston's Museum of Science, was putting finishing touches on the detailed National Geographic map of Mount Everest. He and Swissair Photo Surveys of Zurich carefully evaluated the new GPS data, fully aware of its fatal flaw: GPS methods can precisely fix the distance from the summit snow to the center of the earth, but tell us nothing about the geoid's elevation. Any new estimate of Everest's height would still have to rely on earlier assumptions for the geoid. The National Geographic map went to press with 8,848 meters (29,028 feet) printed on the summit.

Washburn is nearly certain there is far more snow on the summit than the Italians and Chinese estimated—possibly 15 to 20 feet. In the late spring of 1992, he arranged for guides Pete Athans and Vern Tejas to drive a long ice screw into the summit snow, to measure its depth and to support a laser prism. Washburn's probe augured through a thick layer of ice about three feet below the surface—presumably where the Chinese and Italian probes had stopped—before they ran out of pipe sections, at 7.5 feet.

In September 1992, an Italian/Chinese team drove several probes into the snow summit to a depth of 8.4 feet, and issued a new height for Everest: 29,023 feet. Could these probes have encountered yet another, even deeper layer of ice? In May 1997, Washburn hoped to answer this with an ice-penetrating radar probe, which would map the actual rock summit, but climbers that season were unable to carry the bulky device.

Although the height of the snow summit and the thickness of the snow can be measured to an accuracy of about one inch, defining the geoid remains in doubt by as much as three feet. Because this can be better determined only by taking numerous measurements between Tibet and the world's oceans, it will be several years before scientists can be confident of Everest's exact height. During the wait, the summit may actually rise a few inches as well, based on our measurements of uplift.

Breashears accepted, and he began by providing specifications for a modified IMAX camera. He soon found out what many had learned the hard way: In IMAX, nothing is easy.

"Weight will be one challenge," Breashears said with visible concern. The risks of climbing above 26,000 feet were well documented—and frightening, as he knew from earlier expeditions. He had escorted one blind and exhausted climber down the Southeast Ridge, and had recovered bodies from all over the mountain. Most of these climbers were victims, at least indirectly, of the effects of altitude. On Everest, the margin of safety is extremely small.

Greg MacGillivray and David Breashears wanted to produce a film that would educate as it entertained. To identify and explore the unique issues that relate to Mount Everest, they carefully assembled a team of ten academic advisers. These advisers, all longtime observers of the Himalaya, portrayed an unusual and dynamic mountain: Geologists described Everest not as a static geological monument but as a mountain in motion; meteorologists explained that the Himalaya and its associated plateau are thought to influence much of the world's weather patterns; physiologists observed that Everest's summit is at the very limit of man's ability to survive; and anthropologists portrayed the rich culture of

Veteran climber and deputy team leader Ed Viesturs has summited Everest four times. Currently, "Steady Ed" is well on his way to scaling the world's 14 tallest peaks—all without bottled oxygen.

the Sherpas, a resourceful people who have thrived in the shadow of these mountains for more than four centuries.

Over a period of a year, an experienced team was compiled, consisting of four climbers, four filmmakers, two Base Camp support staff, twenty-four Sherpas, and three on-site advisers and journalists—a total of 37 people on the mountain.

TO KATHMANDU

Prior to 1950, Everest expeditions approached the mountain through Tibet. Nepal's complementary barriers of topography and isolationist policy had effectively insulated it from the rest of the world. But in that year, Nepal began to open its borders to the outside world. For climbers, the gates to a tantalizing, magical, and daunting kingdom had been unlocked. Eight of the world's ten highest mountains are located within Nepal (or on the border with Tibet or Sikkim), and until the 1950s none of them had been climbed.

In early March 1996, the *Everest* film expedition members departed their countries for Kathmandu, Nepal's capital, a scenic destination from any approach.

As a plane begins its descent into the city, half the Nepal Himalaya are visible. From Dhaulagiri to Everest, the world's most imposing peaks rise like an apparition. Mountain slopes graced with snow and forested high valleys beckon seductively. The plane crosses a pass and the city appears, surrounded by corn and rice terraces that are sculpted on the rim of the Kathmandu valley like topographic map contours.

From the taxi window, it is clear that this ancient city is embracing the transition from the 12th century to the 21st. But the pavement and neon seem a temporary and superficial dressing, as if the city's urban cows, wandering sadhus, centuries-old temples, and stone deities will momentarily rise up and sweep aside the trappings of the modern age.

Appearing a bit unusual, the team members converged at the Yak and Yeti Hotel. "What country is your expedition from?" one hotel guest asked. Spain, Japan, Austria, the U.K., India, Nepal, and the U.S. to start with. Stacked on the reception desk was a colorful pile of dog-eared passports.

CLIMBERS, FILMMAKERS, AND ADVISERS

David Breashears knew that in selecting climbers, experience and compatibility would be critical; the team would be spending at least two and a half months together. He and Greg MacGillivray wanted an international team as well and to offer a shot at Everest for expert climbers who might not otherwise have the opportunity.

Assistant cameraman Robert Schauer of Graz, Austria, made numerous ascents and adventure films before joining the crew. His work, he says, "is fun but also frustrating. I try to develop a relationship with the camera, so that operating it becomes automatic."

FOLLOWING PAGES:
Clearly above the crowd, a distant Everest edges out its nearest neighbors: Lhotse, Lhotse Shar, and a host of other cloud-wrapped peaks.

For deputy leader, Breashears chose American mountaineer Ed Viesturs, a three-time Everest summiter who was keen to climb it again, his third time without supplemental oxygen.

After graduating from the University of Washington in 1981, Ed attended Washington State University's veterinary school and guided for RMI, a guide service on Mount Rainier. In 12 years, he summited Rainier 187 times. He also climbed Denali and Aconcagua before being invited to Everest in 1987. Friends have nicknamed him "Steady Ed" for his consistency and professionalism. Viesturs is one of five people to climb the world's six highest peaks without bottled oxygen, and the only American. He has embarked on a quest he calls "Endeavor 8000" to summit the world's 14 highest peaks, those over 8,000 meters (26,250 feet), without supplemental oxygen. He has climbed nine of them thus far. "This is a personal goal for me," Ed explained, "it's not for fame."

"One of the joys of being with Ed is knowing you are in the presence of a superior being," Breashears summarized. "His stamina and reliability are phenomenal."

Ed's wife Paula Viesturs had worked in Base Camp on a previous expedition, and Breashears selected her to be Base Camp manager.

A hiker and beginning climber, she would provide logistical support while the climbers were on the mountain.

Breashears next chose Araceli Segarra, of Lleida, Catalonia, whom he had met a year earlier on the north side of Everest. A professional physiotherapist, Araceli is a versatile ice, rock, and alpine climber. She climbed the South Face of Shisha Pangma, 26,291 feet, alpine style in 1992, and in 1995 she attempted the North Face of Everest, reaching 25,591 feet. If successful, she would be the first Catalan—and Spanish—woman to climb Mount Everest.

Araceli came to climbing through spelunking. Having trained for years in Spain's vast limestone cave systems, she brought an abundance of enthusiasm and optimism to the team. "Mainly, I like to climb with friends and to enjoy, especially when there's a nice route. When I climb a rock wall I dance and connect movements without thinking or breathing the words 'I'm falling, I'm falling.' And with friends—there is nothing better."

Climbing Leader Jamling Tenzing Norgay, of Darjiling, West Bengal, India, is an experienced mountaineer and expedition organizer. This would be his first Everest attempt. His father, Tenzing Norgay, reached the summit of Mount Everest with Sir Edmund Hillary on the historic first ascent in 1953. A devout Buddhist like his father, Jamling deeply respects the deities that reside on and around Everest.

Spelunker and alpinist Araceli Segarra, 27, added youthful energy and international flavor to the team roster. Born in Catalonia, she says of rock climbing, "There is nothing better."

"Each year on the anniversary of my father's climb, the 29th of May, at least one member of my family places a silk *kata* blessing scarf on the scroll painting of Miyolangsangma, the goddess of Everest, that we keep on our altar. It was to her that my father expressed gratitude when he reached the summit.

"I'm hoping that the culture of the Sherpas will be viewed more widely as a result of the film of this expedition," he added.

Junko Tabei was the first woman to summit Everest, and in the 20 years since her historic ascent no other Japanese woman has climbed it. Breashears met Sumiyo Tsuzuki in 1990 on the north side of Everest. She had been on Everest twice before, and had climbed to 23,000 feet on Everest's North Col in 1995. Sumiyo would be documenting the expedition on video.

"To go on an expedition and climb you need high concentration, which we do not have in city life," Sumiyo said with a shy, girlish smile. "That, I enjoy."

To help muscle around the streamlined IMAX camera—and for an extra head to think in conditions of oxygen starvation—David would need to work with another accomplished climbing cinematographer. Robert Schauer of Graz, Austria, assistant cameraman for the film team, is David's European mirror image. "I had read about Robert when I was younger," Breashears said, "and it's an honor for me to climb and to film with him."

Schauer has produced or co-produced 12 mountain and adventure films and documentaries, and has climbed 5 of the world's 14 8,000-meter peaks. He was the first Austrian to reach Everest's summit, 18 years earlier.

"David has shown great sensitivity in selecting the team," Robert noted. "We are already working together well, and our humor and spirit are flowing."

A few days after the climbing and film teams arrived, a seasoned Nepal hand dragged several battered bags into the Yak and Yeti. They belonged to Roger Bilham, Professor of Geology at the University of Colorado in Boulder and adviser to the film. Two of the suitcases were filled with tools,

Son of a legend, Jamling Norgay faces Everest as he prays to its resident goddess.
In 1953 Jamling's father, Tenzing, and Edmund Hillary made the historic first ascent of the mountain.

transistors, super glue, batteries, a laptop computer, and some heavy wire that looked suspiciously like bent coat hangers.

"Roger knows a great deal about everything," David said proudly. Ed referred to the transplanted British national as "our team geophysicist." Roger had already been working with the government of Nepal on establishing a network of Global Positioning System (GPS) satellite receivers to measure movements of the tectonic plates beneath the Himalaya. He would be taking GPS readings along the approach route—which partly follows the dynamic junction of the Eurasian and Indian landmasses. He would also oversee the placement of the world's highest weather station, at 26,800 feet

IMAX ON EVEREST

THE DEVELOPMENT AND TESTING OF A MODIFIED LARGE FORMAT CAMERA

I figured it would be impossible to struggle to the summit of Everest with a standard, 85-pound IMAX camera—and inconceivable that a smaller, lightweight version would be sturdy enough to withstand Everest's harsh conditions," said David Breashears.

For more than a year, Kevin Kowalchuk, Gord Harris, and their technical team in the IMAX Corporation's engineering department worked to build such a camera. After conferring with Breashears, they determined the specifications needed for an IMAX camera to actually work on Everest:

• The weight of the camera body could not exceed 26 pounds. Even when climbing with supplemental oxygen, hypoxia would make it extremely difficult for a climber to carry load heavier than this above 26,000 feet.

• The camera would have to withstand minus 40°F temperatures for 24 hours and then operate reliably with the flip of a switch.

• Large, accessible knobs and lens mounts would be needed to allow an exhausted camera operator to film with potentially impaired motor and thinking skills and hands stiff from the cold or covered with mittens.

At IMAX Corporation headquarters in Toronto, the engineering team began with an IMAX Mark II camera model, modified the lightweight and durable magnesium body, then simplified the electronics and removed the 8-pound flywheel.

"We turned a '96 Chevy into a '56 Chevy," said cameraman and technical supervisor Brad Ohlund, "while reducing its weight by half." At exactly 25 pounds, the Everest model weighed in at less than half the Mark II's normal weight.

"At minus 40°F, the outside dimensions of the camera will shrink at least 1/10,000 of an inch," Kowalchuk explained, "This is significant, because the aluminum and magnesium exterior parts expand and contract at different rates from the steel interior parts, which could jam the camera mechanism. We made allowances for it."

Extreme cold congeals lubri-cants and makes film brittle, so the engineers built in a hand crank for loosening up the system in the most extreme weather, and used a lubricant that remains viscous down to minus 100°F. Conventional batteries fail in the extreme cold, so the camera needed to be powered by a nonrechargeable lithium battery pack.

When fully loaded with lens and 500-foot film magazine, the camera package weighed in at 48 pounds. The film alone is heavy, at nearly 5 pounds per magazine, and a lot of it is needed: Film hustles through the IMAX camera at nearly 5.6 feet per second—four times the speed of 35-mm. A 500-foot magazine would last only 90 seconds.

At 15 perforations per frame, IMAX film is the largest commercial film format. In terms of surface area, it is 10 times the size of 35-mm feature film format, and 3 times larger than 70-mm. The film runs through the camera horizontally, as opposed to vertically as in other film cameras.

After testing the camera system in a blizzard on a New Hampshire mountaintop, Breashears and Kowalchuk rented a cold test chamber used by the defense industry.

"When we stepped in, it was minus 50°F," Breashears said, "giving us an opportunity to test our expedition's down clothing, too, which we were wearing." They loaded film barehanded, plugged in the battery, and turned the camera on. "It performed flawlessly every time. That was the first moment I thought to myself: we can really do this."

Breashears and Kowalchuk also field-tested the camera in Nepal. With 40 porters they trekked over 160 miles, ascending and descending more than 80,000 vertical feet. They left the camera outside overnight, on rocky outcrops, with only a nylon cover for protection. "In the mornings, I'd turn the camera on," Breashears said. "In 27 days not a single scene was lost to camera or battery malfunction."

"We've overcome the technical problems," Breashears concluded. "Now all we needed to do was to get the thing to the summit."

David Breashears films near Camp I with the lightweight, IMAX camera he helped redesign.

on Everest, and take a GPS reading there as well.

Another British national joining the team as correspondent was historian Audrey Salkeld. She had written a book about George Mallory and Sandy Irvine, the Everest explorers who had disappeared in 1924, either en route to the summit or while descending. Though not a climber herself, Audrey had been to the north side of Everest with David Breashears in 1986.

Steve Judson, one of the more experienced writers and directors of large-format films, had been on numerous shoots, but none quite as remote or logistically difficult as Everest. As co-writer (with Tim Cahill) and editor of the film, Steve would be working with photogaphic and technical consultant Brad Ohlund, who has 20 years of experience as a second-unit director of photography.

Steve and Brad would travel to Base Camp. En route, they would help David and Robert develop their large-format cinematography skills. Once the camera went above Base Camp, David and Robert would be on their own.

The team knew that virtually no ascent of Everest had begun without the indigenous people of the Everest region, the Sherpas. Most of the 20 Sherpas hired for the film expedition had flown or hiked to Kathmandu from Khumbu, the valleys that enclose the approach route to Base Camp. They brought with them, as always, their customary good cheer and playful humor.

As determined as she is upbeat, beaming team member and videographer Sumiyo Tsuzuki of Japan made it all the way to Everest's South Col—despite two painfully cracked ribs, caused by a severe cough.

Wongchu Sherpa, the sirdar (head man) in charge of the Sherpas and Nepal logistics, was among the most entertaining members of the team. He seemed to revel in the constant battle with bureaucrats over the frustrating permit process, and over which his joking banter and quick-wittedness usually prevailed.

David and Wongchu were on Everest together in 1986. Wongchu himself had climbed Everest from the north in 1995, and had acted as sirdar for several expeditions. "I now hire the climbing Sherpas who hired me— we joke about this," Wongchu said, smiling.

WHY CLIMB?

For climbing Sherpas, climbing is a job, not something they would undertake for sport. A climbing Sherpa can make $1,500 to $2,500 a season, as opposed to the country's per capita income of $175 a year. But only the top sirdars and those with access to investment money own houses and businesses in the capital— where land prices exceed those of many wealthy American residential communities.

So what motivates the others who come to Everest?

"Maybe they have lots of time and money and don't know what to do with it," one Sherpa offered. Paradoxically, though religion forms the fabric of the Sherpas' lives, it may be the foreigners who bring a more introspective, philosophical approach to the mountain.

"Willi Unsoeld, my articulate climbing companion on the West Ridge in 1963, saw in Everest the values of dreaming, striving, risk, going beyond oneself, and caring for our earth and our relationship to it," wrote Tom Hornbein.

Breashears agrees. "Climbing Everest is about the deprivations, the challenge, the sheer physical beauty, the movement and rhythm. And it's partly about risk. You learn about yourself, about what happens when you abandon comfort and warmth and a daily routine, the tyranny of the urgent. You learn how you perform and how you handle a situation that may be life threatening. There's a reward for your effort and a lot of fatigue, too, but I even like the fatigue. I like to wake up in the morning feeling stronger than I was the day before."

Clearly, some climbers are on a quest and set out not so much to conquer a physical obstacle as to attain a new level of understanding of themselves. Climbing a high mountain may be modern man's outlet for the classic hero's struggle codified by Joseph Campbell: approaching, confronting, and then overcoming the weaknesses and demons that haunt us and obstruct us. For this quest, Everest offers the ideal tableau.

Sherpas and other Himalayan Buddhists express this struggle in the form of pilgrimage and in daily rituals. They seek spiritual liberation, but they will settle for gaining merit. The Sherpas understand what the foreign climbers are seeking. They're just not sure that climbing mountains is the best way to find it.

Some climbers who attempt Everest appear to be searching for a sense of power and achievement, for fame and attention. The Sherpas question whether these are proper motivations for being on the mountain. Luck, skill, desire, and money are not sufficient to get one to the top. In order to succeed consistently, the Sherpas say, one's motivation must be pure. They believe Everest and Khumbu's mountains are the domains of gods and goddesses. The mountains exist as much in the realm of the spiritual as they do the physical. Both groups—Sherpas and foreign climbers—share the recognition that humans did not create the Himalaya and that they are beyond man's control.

A CROWD GATHERS

In the Kathmandu market, the team ran across New Zealand guide Rob Hall, who claimed to have escorted 39 climbers to the summit of Everest—more than the total number of people that summited during the 20-year period following the first climb, in 1953. This year, he brought eight clients, two more than he had ever brought to the mountain.

American guide Scott Fischer also arrived with a team of guides and clients. Fischer had been on Everest once before, in 1994 when he summited without oxygen. This year his Seattle-based company, Mountain Madness, had taken on several clients, among them New York adventurer Sandy Hill Pittman, who, if she summited Everest, would have successfully climbed all "Seven Summits," the highest mountain on each continent. Another client was climbing legend Pete Schoening, widely known as a hero for having saved the lives of six team members high on K2, in 1953. If

In step with tradition, Jamling lights butter lamps in the monastery at Khumjung village, in supplication to various dieties. Four decades earlier, after performing similar rites there, his father ultimately stood triumphant (right) atop Everest.

SECOND GENERATION CLIMBING LEADER

BY JAMLING TENZING NORGAY

Born in 1965, Jamling, short for Jambuling Yandak, or "world famed," was the climbing leader for the Everest *Film Expedition. His father, Tenzing Norgay, made the first ascent of Mount Everest with Edmund Hillary, in 1953.*

My father, Tenzing Norgay, was my mentor and role model. When I was young, he took me trekking in the Sikkim Himalaya, where he taught me to climb. At age six I scrambled up a peak with him.

Traditionally, Sherpa sons follow in their fathers' footsteps. But when I told my father of my dream to climb Everest, he said,

"Why? I climbed so my children wouldn't have to—so you can get a good education and not have to carry loads in the mountains, risking your life." That's what he wanted: to give us everything.

Once I started climbing, though, he didn't discourage me. At St. Paul's School in Darjiling I organized rock climbing demonstrations. Climbing was in my blood, it seems.

In 1984, the Indian Everest Expedition put the first Indian woman, Bachendri Pal, on the summit. I was 18 and had wanted to join that expedition. I had hoped to be the youngest person

to climb Everest. I was a pilgrim craving reunion with the mountain that drew me, just as I am now. But my father was sick, and I decided not to go.

I then came to the United States to study at Northland College in Wisconsin, where my father had received an honorary law degree in 1973.

America was exciting and stimulating, but something was missing for me. My father died in 1986, and in the fall of 1992, while my two brothers, one sister and I were living in the U.S., my mother died. Our roots in the Himalaya were about to wither, I felt, so I chose to return.

With the death of my parents, my determination to climb Everest became even stronger. I organized a small expedition in 1993 to commemorate the 40th anniversary of my father's Everest climb. In my heart I knew that reaching the summit was only a matter of time.

I became engaged to my fiancée Soyang, whom I had gone to school with in Darjiling. Our parents were good friends, and they arranged the marriage —sort of. Even in school, I always knew I'd marry her one day. But she didn't always know of my dream of climbing Everest.

Schoening were to reach the summit, he would be the oldest person to climb Everest, at 68.

At the expedition staging area in the crowded center of Kathmandu, the climbers and Sherpas moved bundles of equipment, gas cylinders, duffels, and boxes of food, while Ed and Paula logged each item. Shooing away curious jungle crows that squawked and hopped about, they stood back and scrutinized the gear and the confusion.

"We organized our wedding at the same time we prepared for the expedition," Paula said. "It *was* hectic."

Calculating quantities, locating supplies, and buying and packing the food and equipment became a full-time job for three months. Three tons of gear were shipped from Seattle: 57 food boxes, 30 climbing hardware and tent loads (including over 40 tents and 50 sleeping pads), 5 science cases, 3,000 feet of rope, 75 bottles of oxygen, hundreds of rolls of toilet paper, and 47 tins of Spam. The film gear was shipped separately. Propane, kerosene, petrol and additional food and supplies would be purchased in Kathmandu. Eventually, 250 loads would have to be transported to Base Camp by helicopter, yak, and porter.

The *Everest* Film Expedition, not including the science and film components, would cost nearly three-quarters of a million dollars—a significant location expense. David Breashears was determined to make the summit, but early

Ill-fated commercial guide and expedition leader Scott Fischer of Seattle was one of eight to die in an afternoon storm that blew in on May 10, trapping him and the others near Everest's summit.

on he resolved that he wouldn't allow the high stakes of the expedition to compromise his commitment to safety and caution.

A POPULAR PEAK

The team was fortunate to have picked a year when permits to climb Everest from the Nepal side were available on relatively short notice. Prior to 1978, only one expedition per route was allowed each season, and the waiting list for the South Col route, considered to be the easiest, was more than five years long. But in that year, the Mountaineering Section of Nepal's Ministry of Tourism opened up south-side routes to more than one climbing team—possibly after they observed that multiple teams were climbing from the north side of the mountain, in Tibet.

With only a nominal fee per expedition and few limitations, the number of groups and climbers increased through the 1980s—until the spring season of 1993 when 15 teams, comprising nearly 300 climbers, attempted Everest from the south. The following season, the Ministry of Tourism limited the number of teams to four, while increasing the royalty to $50,000 per expedition of up to five members, plus $10,000 for each additional member, to a maximum of seven.

In the spring of 1996, rather than impose a ceiling on the number of teams, the government increased the fee to $70,000 per seven-member expedition, plus $10,000 for each additional member, to a maximum of 12. As anticipated, the new policy boosted mountaineering revenues. During the spring of 1996, Everest climbing royalties generated more than $800,000, a large sum for a developing country. Many have complained that only a fraction of this amount has been allocated to environmental conservation or mountain safety, and little has trickled down to the local people.

For a client, the going rate is $30,000 to $65,000, and organizing a large expedition can cost over a half million dollars. When Charles Houston, physiology adviser to the film, climbed K2 in 1938, the round-trip cost for the entire expedition was $9,500. The second attempt he led in 1953 cost $35,000.

Attracted by the film's potential to promote tourism, the government of Nepal issued the *Everest* team a permit in the early spring of '96. At the time, Breashears had no idea that his group would be joined on the mountain by 13 other teams, all climbing on the same route and during the same narrow window in May when good weather was expected. The two largest groups alone, those led by Scott Fischer and Rob Hall, would be placing 22 guides and clients and nearly as many Sherpas on a single route up the mountain.

New Zealand commercial outfitter Rob Hall knew Everest's dangers from numerous previous tries and ascents. Yet he also perished in the May 10 storm, running out of oxygen—and energy— while trying to save a client's life.

ASAN TOL AND THE CITY

The team slalomed their way through a maze of people to the very heart of Kathmandu: the busy market area of Asan Tol. Regardless of their destination in the city, everyone intersects at this six-way hub of narrow alleyways, a crossroad of culture and color.

Asan Tol's Annapurna ("Goddess of Abundance of Grain") Temple is appropriately surrounded by burlap sacks of grain, the tops neatly rolled back by merchants for buyers to inspect. Ministers, mendicants, visiting hill people and pilgrims ring the temple's bells and make offerings to the Hindu and Buddhist deities, then jostle their way through rickshaws, overloaded push carts, and breeding bulls that range freely, grazing on vegetable scraps and cardboard. The smells of incense and rotting vegetables meld in a thick, organic vapor.

The self-proclaimed Global Emperor, a Kathmandu temple pundit and eccentric, greeted David Breashears in an alley near the temple and handed him playing card-size pictures of deities of the Hindu pantheon, with his "global messages" inscribed on the back. David listened carefully as he imparted advice that contained, without irony, the two most repeated axioms of life in Nepal: "Oh, no, sir, you can't do that here" and "Here, my good man, *anything* is possible."

Filming in Kathmandu proved both axioms. Steve Judson rapidly adjusted to the chaotic, medieval location with his sense of humor intact. "The first few times I called for 'action,' everybody stopped and looked at me. Finally, someone informed me that the Nepali term *ek chin* means 'wait a minute!' "

The teetering, overbuilt character of the old section of Kathmandu lent a mild foreboding to the proceedings. The valley is situated above a huge earthquake fault that underlies most of the Himalaya, and some of the valley's residents are old enough to remember when the city was devastated by the Great Bihar Earthquake of 1934, a year they still count time from. Some seismologists believe that the

Last chance for fresh fruits, vegetables, and other supplies, Kathmandu's colorful Asan Tol market has become a mandatory stop for Everest expeditions.

PAGES 42-43:

The Kathmandu suburb of Bhaktapur nestles amid a Himalayan backdrop, crowned by the 22,930-foot peak of Dorje Lhakpa.

next "great" earthquake is overdue; when it occurs, it will affect a city population that has grown tenfold.

One concerned citizen is Kathmandu building engineer Hemant Aryal. He recognizes that the cantilever design of many buildings has created welcome space in the city's cramped core. But because the overhangs of the cantilevers can be no wider than three feet, the owners tend to construct the walls with bricks stacked on their sides. In addition, the quality of cement mortar is "variable."

"The bricks can be easily dislodged by even the smallest horizontal thrust," Mr. Aryal writes. "In a quake, as people run out of their houses to escape, they will be greeted by a rain of bricks."

THE VALLEY OF KATHMANDU

Legend says that Buddhist Lord Manjushri spied a crystal clear lake in the middle of the Himalaya and drained it with one pass of his sword, creating a broad, pastoral valley with deep lacustrine soils.

The Kathmandu valley is lush, mysterious, and chaotic. It is the capital of a halting but spirited democracy, a mandala-shaped holy power spot, a staging point for expeditions, a way point for traders. Even the skeptical can feel the valley's spirit: a frenetic mix of incense and vermilion powder, tea shops and temple bells, monks and monkeys—and erotic relief carvings on the roofs struts of temples. Swarms of paper kites chase purple martins, and three-dimensional clouds assume the shapes of demons or deities, depending on your mood.

Jeweled, turbaned, sometimes ragged hill villagers trek through the endless market on a combined pilgrimage and shopping spree, touching passing sacred cows with respect, sharing with foreigners mutual glances of wonder. Then the whole city stops for a royal motorcade—down a road divided by a white line freshly laid by nine workers who were issued only one paintbrush.

One of Asia's richest cultures sprang from this fertile lake bed. The Newars, the valley's indige-nous people, have clustered their "urban farm" hamlets around sacred sites—and around a soci-ety of self-sufficient clans. Not only farmers, they are also priests, artisans, craftsmen, and traders. Their architecture dominates the valley and has influenced the designs of Buddhist and Hindu monuments throughout Tibet and western China. Buddhism itself sprang from Hinduism, and the two faiths are beautifully reflected in the Newars' unique artistic synthesis and in their elaborate iconography and myths. Of course, Buddha was born in Nepal.

There are more recent influ-ences as well. British Army officers—among the few out-siders allowed to enter Nepal before 1950—describe climbing from ridge top to valley and back to ridge, as guards dis-patched smoke signals to alert the next station of the arrival of a foreigner. Imagine their surprise upon descending into this for-mer lake bed, 15 miles in diameter, to find an occasional car—each one carried in from India by porters, who also car-ried them back for repairs. In 1956, a winding road connected Kathmandu to the outside world for the first time.

Change has come to stay in Kathmandu, in the form of satellite television, cell phones, pollution, growth, and even more roads. But the culture, the clouds, the fantasy, and the palpable image of Lord Manjushri's sword may continue to live, as mythology measures it, "a Milky Way of years, times thirty-three."

Shade trees and terraced croplands accent the lush, pastoral Kathmandu valley, ancestral home of the Newars, a people renowned for their rich cultural heritage and artistry.

ROOFTOP OF THE WORLD

Marking the border between Tibet and Nepal, the world's tallest peak honors
Sir George Everest, Surveyor General of India from 1830 to 1843.

Additionally, Kathmandu's narrow streets are festooned with a labyrinth of frayed electrical wires that will increase the existing fire hazard from cooking fires. And after a quake, fire trucks won't be able to negotiate the narrow, brick-clogged streets.

Perhaps prayer will help. Though Kathmandu is virtually choked with motor vehicles and factories, the valley is still dotted with ancient power spots—geomantic focal points of spiritual energy that valley residents believe confer blessings on those who frequent them.

The team members took a break from their stockpiling of supplies to visit Swayambhunath, the monkey temple. For Jamling, Sumiyo and the Sherpas—all Buddhists—this was a chance to prepare in another way: by making offerings to the deities of this holy site.

The stupa of Swayambhunath, an ancient hilltop shrine, is sacred to both the Hindus, who entered Nepal from the south centuries ago, and to the Buddhists, who came from the north. Swayambhunath means "self-arisen," and legend says that the hill on which the stupa is located spontaneously emerged from the lake bed that is now the Kath-mandu valley.

Joining the daily stream of supplicants, Jamling and Sumiyo ascended the 365 stairs to the stupa. They prayed and threw grain from bamboo trays to the hundreds of

While monkeys cavort and pigeons wheel, a young Buddhist monk at Swayambhunath temple in Kathmandu rings a bell to signal the pantheon of Hindu and Buddhist deities.

pigeons and rhesus monkeys that long ago claimed Swayambhunath as their home. These offerings of life bring *sonam,* or spiritual merit, which contributes to a more favorable rebirth.

As they circumambulated the stupa dome, they spun prayer wheels for blessings and good fortune. It is said that the copper repoussé wheels, packed with prayers and mantras block-printed on hundreds of scrolled rice paper folios, release the written invocations heavenward with each revolution. A handful of foreign Buddhists, mixing easily with the locals, also chanted and spun.

The team frantically tried to fit all the color and action into the

widest angle lens, while monkeys eyed the film gear. David reminded Brad Ohlund and Robert Schauer not to leave lenses or equipment unattended, lest they be lifted by these acquisitive rogues.

Jamling paused to purchase a long roll of colorful prayer flags that he planned to unfurl on the summit—after having them blessed by a high lama.

25,000 BUTTER LAMPS

Jamling was saving some of his offerings for the stupa of Bodhnath, the nerve center and soul of a mostly Tibetan community on the east side of Kathmandu.

The stupa's stone skirt has been worn by countless circuits of murmuring, faithful Buddhists, the soles of their buffalo-hide boots scuffing through a thin layer of dust. Motorized traffic, now blocked from entry, growls and shrieks impatiently outside the gates.

For the success of the expedition, but mainly for its protection and safety, Jamling sponsored the lighting of 25,000 butter lamps at the stupa. For most of the day on March 17, Audrey Salkeld and Paula Viesturs worked with 30 of the team Sherpas, twisting cotton wicks and rustling up shallow pottery cups that would function as lamps. By late afternoon, they had arranged them along the outline of the three main tiers of the mandala-shaped stupa.

The Sherpas—and hundreds of devout

At ease beneath a fluttering canopy of prayer flags, a resident rhesus at Swayambhunath underscores the shrine's common nickname: monkey temple.

bystanders, mostly Tibetan—gathered to light the rows of golden lamps. They filled the clay receptacles with their own melted butter, poured from Chinese vacuum flasks. To forfeit consumption of a valuable commodity such as butter, Jamling noted, demonstrates that one is willing to nourish the gods before oneself.

The angled surface of the terraces became slick with spilled butter, and the lamps ignited with difficulty. When the last one was lit, the climbers, filmmakers, Sherpas, and support staff climbed a narrow stairway to the roof of an adjacent building. Finally, the entire team had come together in one place. They hugged and filmed as the golden light infused in them a sense of calm, good fortune, warmth and equanimity, if not a feeling of proximity to the gods. Beyond the stupa, the lights of landing airplanes winked at Venus, which sat placidly above the sunset.

Even a small breeze can make it impossible to light the lamps or keep them lit. For two nights before the ceremony, Steve and David had scouted the stupa and were twice reminded by the caretakers that it was too windy to consider a butter lamp ceremony. Before the team gathered on the rooftop, a Buddhist nun had approached Jamling with a broad smile and bowed her head slightly in gratitude. She said that although the afternoon had been quite windy, it was propitious that the wind stopped. The gods were looking favorably on Jamling's and the expedition's offering.

For Jamling, the nun's words were comforting. His wife, Soyang, was not entirely pleased with his Everest plans, and she remained quietly worried.

"I asked Soyang to give me one chance to climb it, so that we wouldn't have regrets later," he said. "She agreed, but on one condition: that I consult her family's guru, a Tibetan lama, and request a *mo,* a divination. Geshé Rimpoche is an incarnate lama and is known for his accurate mos. Even the staff of embassies in Kathmandu go to him for advice." Jamling paused. "I was afraid he'd tell me not to go."

At Geshé Rimpoche's monastery, Jamling climbed several flights of creaking wooden stairs and entered a humble room, barely large enough to hold a table, bed, and two or three visitors. Rimpoche, swathed in maroon and brocade robes, was sitting cross-legged on the bed. He motioned Jamling to sit, and called his assistant to bring tea. He then consulted his rosary. When he looked up from the string of 108 beads, he said, "Go ahead: you'll be successful, it looks favorable." He smiled, aware of Jamling's strong desire to climb the mountain.

Jamling asked Rimpoche for a *puja,* a propitiatory ritual, as well. At Soyang's family house in Kathmandu, Rimpoche performed a long-life ceremony and studied the Tibetan almanac to prescribe the most opportune day for him to depart the house. Jamling followed his words precisely, unwilling to chance anything.

The morning of the team's departure, Geshé Rimpoche gave Jamling some powder and blessed grains of sand to sprinkle around dangerous places—wherever he might feel afraid, such as avalanche-prone areas and the Khumbu Icefall. He also gave Jamling a tiny cloth bag containing hair and other relics of high lamas, to place on the summit. Most importantly, Rimpoche presented him with a

protective *sung-wa* amulet, a piece of handmade paper that is inscribed with astrological designs and religious symbols, then folded precisely and bound neatly by a crosshatch pattern of colored threads. Jamling wrapped the amulet in plastic, to protect it from sweat and dirt, and had it sewn into a yellow brocade bag, along with some "long life" pills.

Jamling was greatly relieved. Before leaving Darjiling, he had consulted Chatrul Rimpoche—his own family's guru—who had also done a mo. Jamling hadn't mentioned that mo's disconcerting results to Soyang.

"Chatrul Rimpoche didn't tell me not to go, but he said Everest looked a little uncertain

Copper prayer wheels adorn Swayambhunath temple, each incorporating hundreds of written prayers and mantras. By rapidly rotating the wheels, say believers, invocations are sent heavenward.

this year. I could tell that he saw something difficult in the prayer beads—danger, it seemed. He said nothing more, other than to be careful and that he'd pray for us."

Even Jamling was aware that divinations, though often accurate, could be interpreted in several ways. More importantly, he must maintain pure intentions on the mountain, the clarity of purpose and dedication that Buddhists refer to as "right motivation."

TO KHUMBU

The team was prepared, officially, logistically, physically, and spiritually. Only last minute details remained: constructing a frame for

THE GREAT STUPA OF BODHNATH

BY KEITH DOWMAN

Over the centuries, innumerable Buddhist pilgrims have descended from the Tibetan plateau and high Himalayan valleys to worship at the Great Stupa of Bodhnath.

For these pilgrims, the stupa is a wish-fulfilling jewel at the heart of the mandala that is the Kathmandu valley. To the valley's inhabitants the stupa is the greatest of their Buddhist monuments, its origins lost in the mists of antiquity. Travelers and traders from Kathmandu beginning their dangerous journey to Tibet would stop here for a blessing before setting out to cross the bandit-infested mountains. Today, pilgrims come from all over the Buddhist world for the stupa's magical blessings.

The Great Stupa is an archi-

Tibetan names for this miraculous stupa is "the Fulfiller of All Prayers."

From the air the stupa itself is like the Kathmandu valley, a Buddha mandala. Its parts symbolize aspects of Buddhahood. The three geometric terraces represent the foundation of meditation. At the base of the dome is a three-foot-high circular terrace. This supports 108 niches containing fine sculptured stone idols of the Nyingma sect's pantheon. The splendid dome itself, approximately 120 feet in diameter, symbolizes the womb of emptiness in which all things arise. It is coated by a thick layer of whitewash and decorated with the form of a double lotus— achieved by splashing broad semicircles of saffron water on the whitewash.

The box on top of the dome, oriented toward the eastern point of northernmost rising of the full moon, has the all-seeing eyes of the guardian Buddha painted on each of its four sides. The gilded thirteen steps of the spire represent the ten levels of compassion and the three levels of tantric awareness on the path to Buddhahood. The umbrella indicates the sovereign sanctity of Buddhamind, and the crowning jewel its wish-fulfilling capacity.

Most pilgrims circumambulate the stupa muttering the incantations of tantric Buddhism, while some prostrate, completing their circuits by outstretched body length. Many light butter lamps "to irradiate the darkness of unknowing," and on auspicious days the entire stupa is lit by as many as 108,000 lamps to beseech the Buddhas to answer a special prayer.

Whatever their mode of worship, most Himalayan pilgrims conclude the day in one of the drinking establishments located in the all-seeing stupa's vicinity. In this stupa's nonjudgmental view, there is no discrepancy between devotion and sensory pleasure.

Keith Dowman, a Buddhist scholar, lives in Kathmandu.

tectural representation of the enlightened mind of the Buddhas. Its sanctity is derived from the relics of an early historical Buddha and a saintly Nepalese king that are enshrined in the dome, and by the prayers of some of Tibet's great Buddhist masters. According to the ancient Legend of the Great Stupa, this monument is "the supreme receptacle of the mind of the Buddhas of the past, present and future; to both men and gods, whatever supplication is made and whatever prayer is offered, all wishes will be granted and even supreme realization and spiritual power may be obtained…." Indeed, one of the

Roger's solar panels, signing post-cards, and changing dollars into bags of rupees for paying porters and yak drivers. At least they wouldn't have to carry baskets of heavy coins, as Charles Houston did with the first party of foreigners to enter Khumbu, in 1950. The Sherpas have long since overcome their distrust of paper money and will now even accept traveler's checks.

On March 20, the team assembled at the Kathmandu airport. and climbed into a hulking Russian Mi-17 helicopter, a relic from the Afghan War, and squeezed onto fold-down benches along the windows. As the Mi-17 noisily lifted off, the team's whoops and chants nearly drowned out the rotor blades. Jangbu, a climbing Sherpa and Wongchu's right-hand man, shook his head skeptically. "Too much vibration."

The team was informed that the helicopters had been used as air taxis for years in the former Soviet Union. As the whirling blades strobed the faces of the group Araceli looked around nervously. "For how many years?" her worried face seemed to ask. She noticed the paratroopers' clip line leading to the door. The panel behind it was worn shiny from apparent decades of use.

The chopper crossed the end of the runway and its shadow leapt out over the rice paddies. It rolled left, to the east, headed for the airstrip of Lukla in the region of Solu-Khumbu, the northern part of which is termed "Khumbu."

Masked dancers in Kathmandu reflect the city's cultural diversity. Time and the seasons are marked by festivals held throughout the year.

As they rose above the diffuse fog blanketing the valley, the Himalaya emerged. Their crystalline forms pushed against a navy blue sky as if visibly being thrust upward by the colliding landmasses of India and Eurasia. Like the Himalayan balance of orogeny and erosion, the helicopter's powerful lifting forces were just barely overcoming the relentless pull of gravity.

Because the cabin was unpressurized, the pilots wouldn't fly high enough to avoid the chaotic winds that are channelled skyward by the rumpled foothills. The helicopter leveled out at 11,000 feet, barely higher than its destination, and passed Gauri Sankar. Though this 23,442-foot sacred mountain has been climbed, the Hindus say that Shiva, the God of Destruction, occupies the top of its craggy throne. His consort Gauri sits on the nearby south summit.

One of the mountains most revered by Sherpa Buddhists, Chomolungma—commonly translated as "Mother Goddess of the World" or "Maiden of the Wind" and known to the Nepalese as Sagarmatha—emerged on the northeast horizon. The mountain looked miniature from this distance, but a distinctive white plume blew from the summit like a *kata* blessing scarf. Mount Everest. Jangbu said he wondered why anyone would want to rename a mountain as sacred and majestic as Chomolungma after a human.

In the valley of the Dudh Kosi (Milk River—named for its glacial silt), the chopper turned north and followed the Everest trek route to Lukla. From above, the airstrip looked much steeper than the eight degrees of pitch that has intimidated many pilots, and it dead-ends in a sheer wall of rock. If fixed-wing pilots overshoot it, they'll have a firsthand encounter with the Himalaya itself.

The first airplane landing at Lukla was made by Emil Wick in 1964. David and other veteran climbers still tell stories of this legendary Swiss pilot, who was stationed in Nepal by the Pilatus Porter factory to fly and maintain its high-performance plane. Wick once flew to 32,000 feet, directly over Everest and far above the plane's rated ceiling, while one passenger opened the door for filming. That height could be reached, Wick said, only by seeking out and catching the updrafts that funnel up Everest's Western Cwm to the South Col. After dropping 20,000 feet in a near free fall, Wick sneaked up on the Tengboche monastery—lower than the level of the cupola—and turned to tell his partly thrilled, partly horrified riders, "I want to see if I can spin a few prayer wheels this time." He then jerked the wings to a vertical angle to avoid clipping the tops of two tall prayer flag poles, while people on the ground ran for cover.

The plane's engine was converted from piston to turbine some years ago—a mixed blessing, Emil said, because turbine-powered planes shouldn't be flown upside down.

The team climbed out of the monstrous Russian chopper on unsteady legs. At 9,000 feet, the Lukla air was cool and decidedly thinner. Aromas of juniper fires, yak dung, and turboprop exhaust washed over them, displaced briefly by a light breeze of nostril-tingling glacial air flowing down from Khumbu and Tibet. The smell of Everest.

Moments later, a siren alerted the gawkers and livestock on the runway, and the Mi-17 lifted off in a glorious cloud of dust—to the Sherpas, a wonderful sign of progress.

In the fall of 1964, anthropologist Jim Fisher and an associate completed survey work and land negotiations before grading a sloping meadow to create the Lukla airstrip. It was intended as an access point for school building materials and medical supplies and for the hospital in Khunde, built in 1966 by Sir Edmund Hillary's Himalayan Trust.

A two-week walk had suddenly been reduced to a forty-minute flight. Now, half a kilometer of lodges and shops line the main route that leads out of town. Understandably, the Lukla people are rallying to halt the growing Russian helicopter traffic to the Syangboche airstrip, to the north. Because Syangboche is located closer to Everest, trekkers can now bypass Lukla altogether.

FOLLOWING PAGES:

Pulsing crossroads of colors and cultures, Kathmandu's omnipresent markets promise endless opportunities— not only to shop but to see and be seen.

ON THE TRAIL

Ed, Jamling, and Jangbu oversaw the loading of *zopkios*, male yak-cow crossbreeds that are better adapted than yaks to carrying loads at elevations below 12,000 feet. Slowly, the team moved out

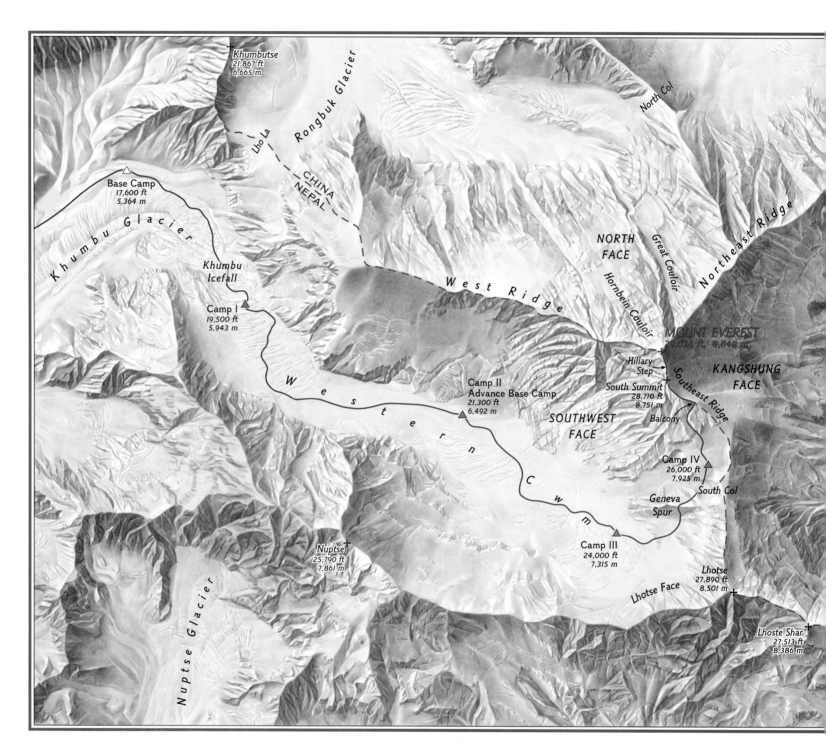

KATHMANDU TO CHOMOLUNGMA

Opting for a 40-minute flight rather than a 2-week overland trek, the Everest *film team helicoptered from the Nepalese capital to Lukla, a village in the Dudh Kosi valley. Yaks then ferried gear and supplies as far as Lobuche village, while porters bore the cargo from there to Base Camp, at the foot of the Khumbu Icefall.*

Rapiu La

1996 *Everest* Expedition,
South Col Route

0   1 mi

0   1 km

N

F a n t a s y   R i d g e

K a n g s h u n g   G l a c i e r

Shartse II
(Peak 38)
24,901 ft
7,590 m

CHINA
NEPAL

Shartse I
24,216 ft
7,381 m

of town and headed up the Dudh Kosi valley. Emerald fields of new wheat and bright yellow blooming mustard formed a quilt along the deep ravine. Lowland porters hunkered beneath oversize loads, their feet the color and texture of elephant hide, with calluses the thickness of sneaker soles. From homes and tea shops perched on terraces along the trail, grandmothers carrying infants in cradles waved the Sherpas in for tea.

Along the Dudh Kosi, white-capped red-starts darted and skimmed about the river rocks, while an ouzel (brown dipper) stood on a rock doing its mysterious deep knee bends, perhaps to gain a better perspective on prey beneath the water. Grandala in their blue-black formal wear hopped about on the pine forest duff above the river. In clearings of potato fields, strikingly black-and-white-striped hoopoes foraged.

The temperate climate put Audrey in mind of a British country garden, and she lingered among the drumstick primulas and tiny gentians carpeting the margins of the path. Yellow cinquefoil poked out of crevices in the trail, and tiny baby-blue butterflies danced in spirals. As the trail went from forest to village, several species of rhododendron appeared, along with blooming, crispy-white bridal magnolias and blushing pink Himalayan cherry.

The team continued north to Phakding, a small community straddling two sides of the Dudh Kosi, connected by a rickety bridge. The gray river banks revealed that the glacial till of the valley bottom had been recently scoured by a large flood.

SILT

Roger kept his eye to the ground and to the cliff sides, seeming to look below their surfaces. The rocks offered evidence of the great continental collision which, he emphasizes, is still occurring.

About 50 million years ago, the Indian continental plate collided with Eurasia, and was forced into it and and beneath it. The compression and uplift from this collision—plus accretions of material scraped off the Indian plate—resulted in one of the reasons Roger and the expedition are here: The Himalaya.

"The most remarkable process occurring here is not so much the growth of the mountains," Roger said, "but their rapid *erosion*. When you observe a series of mountain peaks, you are mostly viewing the absence of material between them."

The rivers carry away the mountains in the form of solutions (chemically dissolved rocks), suspended rocks (mud and sand), and bedload (boulders rolled along the bottom). "We can measure the amount of eroded rock from the rate that reservoirs fill with sediment," Roger explained. "But this is only an approximate gauge, because we believe most erosion occurs during very heavy rainfalls that may happen only once per century—perhaps when we aren't measuring.

"For most of Nepal's river basins, the

Gouged by nature and terraced by man, the treeless ridges of the southern Himalaya comprise a tortured land of monsoon-driven erosion and tectonically spawned uplift and earthquakes.

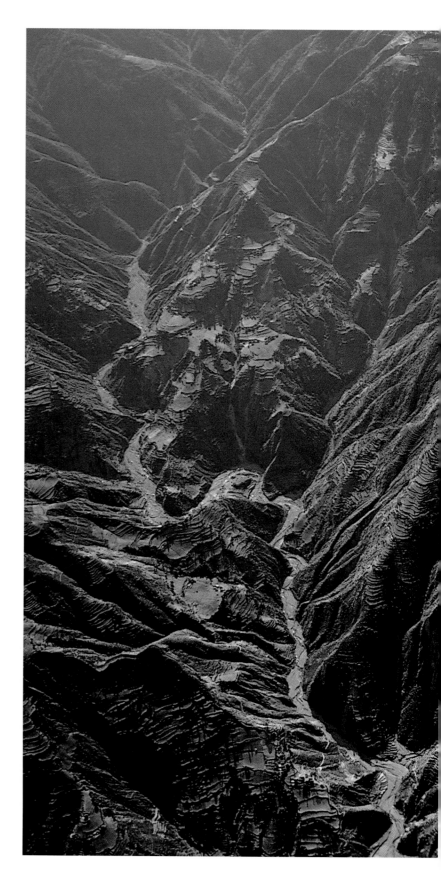

THE GREAT LANGMOCHE FLOOD

The erosive capacity of streams and rivers, and the agitation of a landscape still in motion, are dramatically visible near the lunchstop hamlet of Phakding.

Brian Carson has spent part of a lifetime studying the movement of Himalayan soils, and he summarized the forces eating away at the Himalaya: "Slope, aspect, bedrock type, land use, forest type and condition, wind velocity, and even rainfall erosivity—the small but cumulative explosive impact of raindrops—are all factors in erosion. But another category of erosion is what we call mass wasting—rock failures, landslides, slumps, riverbank cutting and gullying."

One monsoon day in 1985, a single event almost instantly altered parts of the Bhote Kosi and Dudh Kosi riverbeds, which drain the Khumbu. Far upstream, a glacier and its moraine coming out of a side valley blocked or pinched off the larger main valley. The main valley, Langmoche, then filled with water, forming a glacial lake. When the ice melted, the dam formed by the pinching side glacier weakened and collapsed.

The lake overtopped this nat- ural dam, and it quickly eroded the sediments retaining it. The resulting "glacial lake outburst flood" (GLOF) was catastrophic.

Sedimentologists Vuichard and Zimmerman determined that the sudden breaching of the dam was triggered by massive waves generated by a gigantic ice avalanche that collapsed into the lake. About 5 million cubic meters of water were released over a period of a few hours, causing the river to surge to a peak flow estimated at 1,600 cubic meters per second (compared with a normal winter flow of about 2 and a monsoon flow of 30 or more).

The amount of debris swept downstream and carved from the riverbanks was staggering. A hydroelectric plant under construction was destroyed, massive landslides were set off, and numerous houses were swept away, along with 16 bridges. The water and moving rocks shook the windows of Namche Bazar, though it is 2,000 feet above the river, and sent up clouds of water vapor and rock dust. Portions of the trail had to be rerouted around the river damage.

The global recession of glaciers in the past half century has allowed glacial lakes similar to this one to grow in size and number. In the Khumbu, three other lakes—all closer to Everest than the one in the Langmoche valley—have been identified as being at risk, including the large Imja Glacier Lake. It began forming in the early 1960s and now holds over 30 million cubic meters of water, or six times the quantity of the one that fed the massive Langmoche flood on the west side of the park.

Himalayan deities come in beneficial manifestations, and they can be wrathful and destructive. Similarly, the rain and snow they deliver bring fertility to crops—and devastating floods and landslides.

The destructive blast of the Langmoche flood scoured out this Himalayan valley.

overall erosion rate is about 2 millimeters per year. Although we have measured uplift rates of 3 millimeters per year in the foothills and rates of 5-8 millimeters per year in the high mountains, these rates are very localized."

But three centuries of erosion can virtually be reversed during the few moments of rapid uplift that occurs during a great Himalayan earthquake, when plates shift and faults realign.

ONWARD

The team was leaving the area called Pharak and entering Khumbu, the region that since 1976 has also defined the boundaries of Sagarmatha National Park. Schematically, Khumbu is funnel-shaped, and the only access, other than across treacherous glaciated passes over 18,000 feet, is past the park entrance station at Monjo. This hamlet is at the park's lowest point, at 9,200 feet on the Dudh Kosi.

It's all uphill from here. With steady, rocking gaits, climbers and yaks made their way up 2,000 vertical feet toward Namche Bazar. At places along the steep switchbacks the narrow trail drops away 1,000 feet or more. "That's where I learned the first lesson of trekking," Steve said. "Never stand between a yak and a sheer drop-off."

Namche, at 11,300 feet, sits on a terrace far above the river that formed it and is overshadowed by the nearby peak of Kwangde, at 20,298 feet. With Jeff Lowe in 1982, Breashears climbed Kwangde's north face, a difficult new route, alpine style, considered one of the hardest technical climbs ever done in the Himalaya.

Namche is Khumbu's largest village, a market town of a hundred houses and a score of lodges—some of them four stories tall, with 360-degree-view restaurants. There's a bank, offices, shops, and a dental clinic established with aid from the American Himalayan Foundation. Two young western-trained Sherpas are now booked solid doing fillings and other dental work for Khumbu villagers and government staff.

Namche's bustling Saturday market exemplifies free enterprise. Sides of buffalo meat, sacks of grain, and bamboo baskets filled with vegetables and tangerines line the terraces at the edge of town, a porter manning each basket.

A luxurious trek can be outfitted from leftover expedition supplies for sale in Namche shops. After a Spanish expedition, the town was awash in tins of marinated trout from Catalonia. One Sherpa cook nearly gave away a drum full of large glass bottles of Black Sea caviar he had inherited from the retreating Soviet team.

"Fish eggs," the cook scoffed.

On the steep trail that leads out of Namche to Khumjung, the team was buzzed by a squadron of snow pigeons, hugging the valleys and ridges, their sudden whoosh catching them by surprise around a blind corner. The white pigeons banked into a turn and momentarily disappeared, then reappeared heading in the opposite direction, this time preceded by their shadows.

FOLLOWING PAGES:

The bucolic mountain village of Khumjung, just north of Namche Bazar, served as a staging area for the Everest *film team, which stopped here to switch from* zopkios— *cow-yak crossbreeds—to yaks, creatures better suited to higher altitudes.*

CHAPTER TWO

TO BASE CAMP BY YAK TRAIN

"To Miyolangsangma, lady of the vast, unchanging white snows…
accomplish all the work we have entrusted you!"

—FROM THE ABRIDGED PRAYERS FOR THE LOCAL GODS,
TRANSLATED BY TRULSHIG RIMPOCHE AND RICHARD J. KOHN

The mountains enclosing Khumbu are so disorientingly high, and the trails so uneven, that it's easy to lose your balance on the flagstones that line the trail, rocks worn into smooth round shapes by centuries of foot travel. ¶ Taking in Khumbu for the first time, Roger Bilham marveled at the forces that created what he referred to as "this giant museum of meta-morphic petrology." He pulled out his notebook and cataloged the Khumbu's geological riches: "…harder crystals protrude from their matrix of quartz and plagioclase…needles of black epidote, luminous spheres of red garnet and hexagonal bars of blue beryl…." ¶ Above the Syangboche airstrip, the team paused at a large domed stupa, what Sherpas and Tibetans term a *chorten* ("receptacle of offerings"). David Breashears looked up at the painted

Sherpa child gathers boughs of altitude-stunted juniper to burn as incense, a daily ritual for the devout.

eyes of the Buddha Ratna-sambhava and followed their gaze to the south, where tiers of bluish ridges faded into hazy white sky above the Gangetic Plain of India.

"With all the smoke from the industry and cooking fires of India, the surveyors of the Great Trig Survey would have a hell of a time getting a clear line of sight to the Himalaya today," David observed. The haze was confined to the south, but during his test filming in Khumbu a year earlier, an unusual pall threatened to dull even Mount Everest's splendor.

The team continued north twenty minutes to Khumjung, a village of a hundred houses cradled in the shadow of Khumbila, a scraggy 18,900-foot peak that embodies Khumbu's sacred protector god. They entered the town through the *kani*, a covered stone entryway. The meditating and levitating deities pictured on its ceiling panels are meant to deter evil spirits that might attempt to follow those who enter—while symbolically introducing visitors to the sacred nature of their surroundings.

HOSPITALITY IN KHUMJUNG

The team arrived at the home of David's old friends Nima Tenzing and his wife Pema Chamji, where they would stay for several nights. Nima was a kitchen boy for a trekking company when David first met him in 1979, and he later worked as cook and sirdar with David on Kwangde and Everest.

PRECEDING PAGES: Crossing one of the Khumbu's numerous footbridges, each yak bound for Base Camp traditionally carries two 66-pound parcels of gear and supplies. The Everest team required more than 200 animals for its expedition—and it was only one of 30 climbing groups to attempt Everest in 1996.

A section of the stone wall enclosing a fallow potato field had been dismantled to admit the zopkios, which would be unloaded and corralled here. Pema Chamji leaned out from a second-story window, smiling; a breeze blew a genial wave along the cloth valance fitted on the lintel above her. She urged everyone upstairs for "tea," which in late afternoon is a euphemism for *chang*, unclarified rice or barley beer.

Shortly after the team had settled in, a walkie-talkie broke, the first malfunction of electronic equipment. Roger excused himself and promptly burrowed into the alligator clamps, voltmeters and portable oscilloscope he carried with him. But hospitality can't be refused in a Sherpa home, and Pema Chamji interrupted him with a bowl of the nutritious chang. Sherpa tradition dictates that every cup must be refilled at least twice. Roger obliged.

He happily fixed the radio in time for an afternoon jaunt to examine the nearby GPS survey point measured by his research team in previous years. Returning to the house, he became distracted counting tree growth rings in Nima's and Pema Chamji's firewood pile. The oldest were wind-stunted, high-altitude junipers, 120 years old. He had found similar trees in the Karakorum as old as 1,300 years.

"Old trees can tell us about past changes in Himalayan weather," Roger explained. "Wet years produce more growth, and thicker rings. Occasionally, the rings tell us about earthquakes, too: A quake might accelerate hillside creep, which alters the symmetry of that year's growth ring."

The heart of Nima's and Pema Chamji's house is the firewood hearth, used for cooking meals, distilling alcohol, and warming livestock gruel, and for limited heating. Meat, milk and food scraps are not burned in the fire for fear of offending the local gods, Khumbu's guardian spirits, especially Me Lha, the deity that resides in the fire. Chimneys are a recent introduction, but more commonly smoke filters out through the ceilings and windows, depositing a black, shiny resin that greatly extends the longevity of scarce structural timbers and roof shingles.

The large sitting room acted as a communal bedroom for the expedition, and it quickly filled with wet clothes and camera equipment. Nima's own yaks and zopkios were stabled directly below them; Sherpas say the animals' body heat provides some warmth to the floor above.

Liesl Clark, a journalist, had also joined the team. She was covering the expedition for NOVA Online, and had already begun compiling news and science dispatches to be sent out by satellite fax. She opened the window and propped the phone near it.

Khumjung had been wired for

Emblazoned with a Buddhist mantra, trailside mani *stones occur throughout Khumbu. They range in size from pebbles to boulders 30 or 40 feet across.*

PAGES 70-71:

Sherpa elders from the village of Namche Bazar celebrate part of the annual Dumje festival by tossing handfuls of tsampa—*barley flour*—*at a sacred promontory they believe watches over Khumbu.*

electricity only a few months earlier, through underground cables, and many residents now cook with electrical hotplates, rice steamers, and microwave ovens, allowing them to conserve precious firewood. For the expedition, the team would use wood, dried yak dung, electricity, kerosene, and propane to cook the countless meals.

THE FAITH OF THE SHERPAS

Each morning, Nima and Pema Chamji engage in a traditional religious routine. Outside on a rock platform they burn a small branch of juniper as a purifying offering. Then, on their altar in the chapel room, they fill seven offering bowls with fresh water—a pledge of their commitment to achieving Buddhahood. Because the devout must relinquish all attachment to their offerings, water is perfect: It has no intrinsic value and can be given freely. Pema Chamji said that it is best if starlight has shone upon the water bowls since they were last filled.

Both perform three prostrations before the altar in a demonstration of respect and abandonment of pride. They then dip the unbroken tip of a juniper bough in an urn of holy water, and flick it upward and toward the altar three times.

Jamling, too, used their chapel for prayer, and each morning in

Khumjung he prostrated and recited mantras, prayer-like invocations that bring merit, mindfulness and good fortune. Brightly painted in an elaborate mural of deities and demigods, sacred places and hell realms, the walls of the chapel depict a fragment of the map of the Buddhist cosmos. From the chapel's west wall, the goddess Miyolangsangma, who resides on Everest, looked down on Jamling with quiet grace. Riding comfortably on a female tiger, she holds food in her right hand that represents good fortune; in her left she holds a *nyuli,* a mongoose that continually regurgitates precious gems, con-

PRECEDING PAGES:

Stone fences and buildings stud a Sherpa landholding near Thame, a village on the salt trade route that once thrived between Tibet, Nepal, and India.

Jamling presents offerings to the presiding monk in Tengboche monastery (below). Blessing scarves swathe Tengboche's newly enrolled monks (opposite).

ferring wealth. This is the goddess whom Jamling's father had worshiped, the goddess who had granted him passage to her sacred summit in 1953.

In contrast to Miyolangsangma's serenity, the panels around her pulsated with menacing, ferocious demons, forever tempting humans away from the path of awareness and devotion. Jamling stood in quiet, unperturbed prayer.

THE YETI

On the morning of March 24, the team departed for Tengboche. As they switchbacked down to the junction of the Phunki Drangka and Imja Khola Rivers, David

spotted a small flock of Impeyan pheasants, Nepal's iridescent national bird. Directly overhead, Jamling pointed out a lammergeier, or bearded vulture, riding updrafts near the trail, gliding effortlessly on its nine feet of wingspan. He then told Sumiyo to focus much higher above: A V-configuration of bar-headed geese, migrating over the Himalaya, was passing so far over them that it appeared to barely move.

Araceli wondered aloud if they might see a yeti, a Himalayan abominable snowman. In 1974, Lhakpa Drolma, a Sherpa woman from Khumjung, encountered a yeti while herding yaks in a high Khumbu valley. The moment it saw her, she recounted, it attacked, knocked her

Thrice built, twice destroyed—first by earthquake, later by fire—Tengboche Monastery lies roughly midway between Lukla airstrip and Everest Base Camp. Its 35 monks pray, chant, read, and play ceremonial horns (opposite).

out, and left her in a shallow ravine. When she awoke she saw the yeti mutilating her livestock. It killed three of them. Lhakpa Drolma still becomes edgy and fearful when people around her speak of the yeti.

Some believe that the yeti are associated with Khumbu Yu Lha, a class of deities known as a Dharma Protectors, and that they may be emanations of a wrathful cemetery goddess called Dü-tö Lhamo and other deities. But they are not always dangerous, and are sometimes said to be playful. "Yeti" derives from *ya-té*, "man of the high places," though it is actually surpassed in frightfulness by the *mhi-té*, a long-haired, man-sized humanoid that eats people. They say that even a glance from the

76

mhi-té—especially if viewed from below—can cause illness or possibly death. It is best to avoid saying the names of these dreaded creatures. While chatting with some friends at Base Camp a year earlier, Wongchu recalled, someone uttered "yeti" a fraction of a second before a large avalanche broke off the Lho La, a high pass, and rumbled toward the Khumbu Glacier. The group was momentarily paralyzed, then broke into laughter.

In the 1950s, legendary British climber Eric Shipton photographed footprints he claimed were those of a yeti, and the images were published in the *Times* of London. When queried about Shipton's photos three decades later, Sir Edmund Hillary commented, "I'd have to say—having known Eric very, very well—that in all likelihood he tidied up those tracks."

THE TENGBOCHE MONASTERY

From the bottom of the Imja Khola gorge, the team began the 2,000-vertical-foot ascent of a sandy, forested glacial moraine. Halfway up, Jamling stopped and explained to the climbers the meaning of the carved stone tablets lined neatly beside the trail.

"The Buddhist mantra 'Om Mani Padme Hum,' commonly translated as 'Hail to the Jewel in the Lotus,' is carved on these *mani* stones. It's the mantra of Avalokitesvara—Chenrezig in Tibetan—the Bodhisattva of Compassion. We keep mani stones and

stupas—and people, too—to our right as we pass, out of respect."

A few hundred feet farther on, the team emerged from the fir forest and onto a colorful hillside of blooming rhododendrons. They formed a green-and-rose-colored arborway to the Tengboche Monastery, atop the moraine at 12,700 feet.

Tengboche was established in 1923 as the first celibate *gompa,* or monastery, in Khumbu. It now houses 35 monks, a record number that some attribute to the Sherpas' new tourism-derived prosperity: With growing incomes, more families can afford to surrender an able-bodied son to the monastery and support him there. Some monks are enrolled as young as seven, though fewer than half will remain in the monastery their entire lives.

This tranquil place was the site of two disasters. The monastery was destroyed in the Great Bihar Earthquake of 1934, and the abbot died shortly after. The monastery was rebuilt, but in 1989 was ravaged by a fire that destroyed almost all the old texts, carvings, murals, and artifacts.

With the aid of Sherpa carpenters, local patrons of the monastery, and grants from the American Himalayan Foundation, the monks have rebuilt the monastery. The Buddhist concepts of patience and nonattachment have been instrumental in Tengboche's survival.

Gilt colossus of Tengboche, this 15-foot-tall Buddha of the Present dominates the monastery's main assembly hall, where Jamling devoutly sought blessings on behalf of the expedition. A scroll painting at his home in Kathmandu celebrates Miyolangsangma (opposite), tiger-riding goddess who resides on Everest.

The Sherpas draw much of their religious tradition from Rongbuk monastery in Tibet, located at 16,000 feet on the north side of Everest. Rongbuk was destroyed during the Cultural Revolution, which began 15 years after the Chinese occupation of Tibet in 1951. Fortunately, some of Rongbuk's and Tibetan Buddhism's unique and colorful ceremonies endured in Nepal.

Sumiyo and Araceli stood with Jamling on a terrace below the gompa as he chanted quietly in the direction of its imposing stone walls. "I studied and lived in the United States for ten years," Jamling said, " but wanted to return to the Himalaya to learn more about my own culture. I feel I am completing a circle."

The deep drone of horns and drums intensified as the three climbed the stairs and traversed the courtyard to the main assembly hall. They halted outside a massive wooden door and removed their shoes. Jamling stepped over the threshold, walked forward three paces, and paused to view the interior of the hall, his hands placed together in supplication. The low vibration was now as palpable as it was audible.

Directly ahead, a formidable statue of Buddha Sakyamuni, the Buddha of the Present, sat 15 feet tall, his gilded shoulders and head extending through an opening in the second floor. Smaller gilded figures stood in attendance in the

foreground: disciples Shariputra and Mangal-putra, who possess miraculous powers, and the Bodhisattvas Chenrezig and Jambayang. Eight Tatagathas, fully enlightened Buddhas, appeared to levitate within the floor-to-ceiling halo behind the main image. Jamling prostrated three times, lowering himself to the floor and then rising.

Sixteen monks sat in rows facing the main aisle, reciting from long folios opened on low prayer tables. Some rocked slowly as they chanted, while those who had memorized the text continued to pray as they regarded the team with detached interest. As his father had done 43 years earlier, Jamling approached the altar and

Dressed in their most festive clothes, Sherpa women at Tengboche gather to receive blessings and watch dancers perform the Mani Rimdu ceremony, one of the biggest cultural events of the year in Khumbu.

silently presented a long silk kata scarf as an offering of respect to the presiding monk, the Lopon, or Chant Leader. He then handed him the bundle of prayer flags that he hoped to unfurl on the summit.

The Lopon chanted a prayer while touching sacred objects to the bundle. He then poured a handful of blessed barley grains into the bundle's folds: Each kernel is said to contain the qualities and energy of a deity.

For the past three months, the Incarnate Lama had been in strict meditative retreat. When Tenzing Norgay passed through Teng-boche on his way to Everest in 1953, the same lama—then 17 years old—was away in Tibet.

Tenzing was also blessed by the Lopon of Tengboche.

MANI RIMDU

In the courtyard, the team watched the Tengboche monks demonstrate the Mani Rimdu masked dance ceremony of propitiation and blessing, the dramatic closure to an annual rite that lasts over two weeks. Along with its ritual functions, Mani Rimdu introduces the lay community to the history and concepts of Buddhism, amid a socially enjoyable gathering.

Early in the ceremony, a yak is bedecked with katas and anointed with butter in a symbolic offering to the Everest goddess, Miyolang-sangma. The fortunate yak is then released to wander freely through the hills, never to perform work again.

Before the dances begin, monks in tall, tufted hats and maroon robes lead a formal procession, followed by the masked figure of Mhi Tsering (the Man of Long Life), a line of monks playing instruments, the Lopon and his umbrella carrier, and finally patrons and important guests.

In a dozen dances, the colorfully robed and masked dancers depict, among other religious concepts, the conversion of pre-Buddhist demons into defenders of the Buddhist faith. Some dances illuminate the demons and obstacles that create afflictive emotions such as anger, jealousy, lust, and greed.

Masked dancer—a monk—whirls and spins through the monastery courtyard at Tengboche, acting out part of the Mani Rimdu ceremony of propitiation and blessing, which occurs on the October full moon. Other monks also dance or play musical instruments. Performers often represent deities that lie at the heart of human strengths and weaknesses.

The monks' swirling and levitating motions are not merely theater. "Today, despite changes on every hand," Buddhist scholar Richard Kohn points out, "lamas are still the heroes of the Everest region. The monks distribute magic pills to provide sustenance and physical well-being to all who take them. The fearsome deities with whom lama and monastery must deal are paraded before the public. The monks abandon their dull maroon uniforms and don the splendid brocades of tantric magicians, with *mudras* (mystical hand gestures) and magic weapons matching the chaotic forces of the supernatural, threat for threat."

Like any ritual, prayer, or teaching, Mani Rimdu ultimately intends to change the way people think or see, to impart a new vision and awareness. For outsiders as well as Sherpas, simply viewing the ceremony can bring merit, remove obstacles, and allow the fruits of its blessings to accrue.

A NATIONAL PARK

Before 1978, fewer than 3,000 tourists entered Khumbu each year, and most of them trekked to Everest Base Camp. By 1995, the number had grown to over 12,000. The growth of trek tourism—and income for the Sherpas—has led to increased demand for resources, primarily wood for fuel and construction timber, and grass fodder for yaks.

In response to this, the government of Nepal in 1976 designated Khumbu as Sagarmatha National Park (SNP). The park borders the Makalu-Barun National Park and Conservation Area, to the east, and the massive Qomolangma Nature Preserve in Tibet, to the north—forming one of the largest blocks of contiguous protected area in Asia.

The Himalayan tahr, a species of wild goat, can be seen grazing above villages and beside trek routes, along with the rare and diminutive goral, a goat-antelope. Few sightings of snow leopards have been reported in recent years, though Sherpa herders experience some leopard depredation on yaks in their high pastures.

The musk deer, a threatened species, is more common. The male, as large as a medium-size dog, has two long, distinctive canine teeth. Within the park, musk deer are habituated to the proximity of humans, where

Shouldering much of the burden of development, Tibetan women wrestle newly hewn timbers to their village, five days distant. Deforestation (above) is a growing concern in Khumbu and most of Nepal and Tibet.

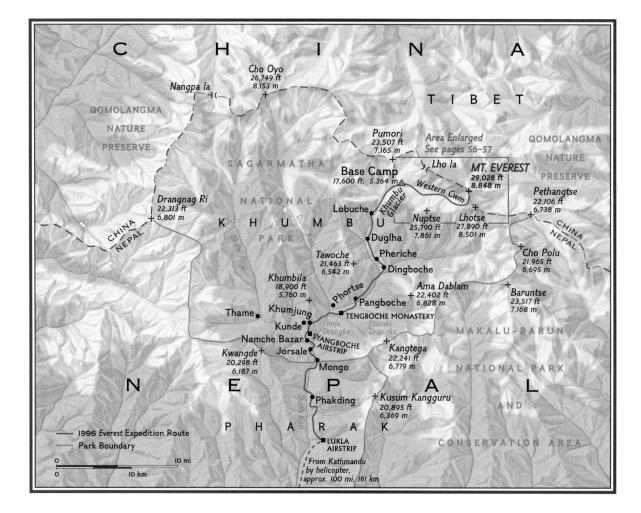

SAGARMATHA NATIONAL PARK

Created by official decree in 1976, Sagarmatha National Park occupies the same three-river watershed that traditionally defines Khumbu. It borders other preserves to the east and the north, all of which harbor increasingly rare large mammals, from snow leopards to musk deer.

predators don't often venture. But there is a flourishing market for the male's musk pod, valued at several times the price of gold. As the species becomes more scarce outside the park, poachers have been lured inside its boundaries. They are often apprehended.

When SNP was created, it wasn't practical to curtail completely the residents' environmental impact. Park managers felt that if

growth could be managed, the Sherpas' relatively sustainable, subsistence lifestyle might complement the park's biodiversty. But trees are being cut faster than they are growing.

More than all other uses of wood combined, burning it for cooking and heating is the most common. Each Sherpa hearth burns two to four metric tons of wood each year, and trekking lodges burn ten times this

THE SHERPAS: HERDERS, FARMERS, TRADERS, AND LAMAS

Tibetan texts relate that about 460 years ago, the Sherpas (literally, "people of the east") migrated from the eastern Tibet province of Kham and settled in the shadow of Chomolungma, Mount Everest. They are a comparatively small ethnic group; about 3,000 of the 35,000 Sherpas in Nepal reside in Khumbu.

Traditionally, the Sherpas are seminomadic herders and subsistence farmers. They raise yaks for milk and butter, hides, and as pack and draft animals, and since the mid-1800s have grown potatoes, thought to have been introduced from British gardens of Darjiling.

During the summer monsoon, Khumbu Sherpas move their yaks and yak crossbreeds to higher elevations, the timing for which is coordinated by elected village guardians. The high *yersa* summer pastures are the last to "green up," and yaks will graze above there, at elevations as high as Everest Base Camp. Yak herd composition has changed in recent years, primarily due to the growth of trek-tourism. The demand for pack animals has led to an increase in the numbers of male yaks and sterile crossbreeds, and a growing scarcity and price of fodder grass.

Sherpas are also traders, and yak trains—somewhat smaller than earlier—still carry grain, butter, buffalo hides, textiles, and other items to Tibet across the Nangpa La pass, at 18,753 feet, and return with salt and wool. The glacier at the top of the pass is tricky. Sherpas say that their more experienced yaks, which they send first, will stop and paw the snow with their hooves to signify when a crevasse lurks beneath a smooth, continuous snowfield. Trade was curtailed in the 1960s with the introduction of Indian salt and the Chinese occupation of Tibet, but to some extent tourism has replaced trade as their primary source of income.

Some have suggested that Khumbu's harsh environment sustains and invigorates the Sherpas' ardent faith. Just as the Himalaya themselves were formed by intersections and accretions of geological material, the religion of the Sherpas and other Himalayan peoples is also accretionary—a blending of shamanism, pre-Buddhism, Bön, Buddhism, and local beliefs.

Buddhism has largely prevailed. The Sherpas are of the Nyingmapa, or unreformed sect of Vajrayana (Tantric) Buddhism, introduced to Tibet in the ninth century by Guru Rimpoche, and in Khumbu much later. When Guru Rimpoche, the great "lotus born" Indian saint known in Sanskrit as Padmasambhava, came to Tibet and Khumbu, he battled the wrathful *srungma* mountain gods of the Bön religion and subdued them, turning them into defenders of the Buddhist faith. He then appointed the Everest goddess Miyolangsangma and four others of the benevolent "Five Long Life Sisters" to look after Khumbu. They now reside on the region's five great mountains, surrounded by other deities such as the protector god Khumbu Yülha.

The Tengboche Lama recounts that Guru Rimpoche forecast that Tibet would be plagued by wars and that devout people would have to flee to sacred Himalayan valleys of refuge known as *bé-yül*. Khumbu was one of four bé-yül that he identified, areas where mystical powers are concentrated and where spirits abound. These valleys should be kept pristine, undefiled by excessive human activity, the Sherpas say. Indeed, sacred sites are scattered over Khumbu's landscape, but few of them are seen by visitors: ancient meditation caves, "self-emanated" impressions on rocks (such as footprints of legendary figures at Tengboche and Pangboche), and curious serpentine intrusions that are evidence of metamorphism—or of the *lu* serpent spirits.

The patron saint of the Khumbu, Lama Sangwa Dorje, born about 350 years ago, announced that important *gompas* would be built at some of these sites. Khumbu's first gompa was built at Pangboche, but it wasn't until 1923 that the Tengboche Monastery was established by Lama Gulu, a reincarnation of Lama Sangwa Dorje's father. Lama Gulu died shortly after the Tengboche monastery was first destroyed, in the 1934 earthquake. His reincarnation is the present high lama of Tengboche, Ngawang Tenzing Zangbu. The gompas of Khumbu are now centers of learning and culture, and form the core of Sherpa spiritual life.

The iconography of the gompas' lavish frescoes represents the Buddhist cosmos, replete with wrathful deities, wish-fulfilling gems, and flying horses bearing bodhisattvas across the heavens. Do these deities, demons, mythical places and magical objects really exist? To Sherpa Buddhists, yes, quite clearly. But do they exist in physical form? Well, yes and no. All physical objects are impermanent and transitory, a creation of the mind, they say—and the mind itself is illusory, proven through Buddhist dialectics to have no inherent, independent existence. For the Sherpas, therefore, Mount Everest is a creation of mind over matter.

amount. In the 1950s and '60s, each expedition consumed more than seven metric tons of fire-wood. But in 1979, the park banned the sale of firewood to trekking groups and expeditions, and halted the burning of outdoor campfires. Expeditions and trekking groups are now required to carry a quantity of kerosene or propane sufficient for the duration of their stay.

Reforestation efforts have helped, and between 1983 and 1995, Sherpas working in three nurseries produced 850,000 seedlings. The saplings must be protected from livestock, which will browse and trample them. With assistance from the Himalayan Trust and the UNESCO World Heritage Fund, 16 fenced plantations, or exclosures, have been constructed.

Conservation has deep roots among the Sherpas. The traditional, locally-appointed forest guardians called *shingi-nawa* are one thriving local institution. From each village, Sherpas select two guardians who control the cutting of green wood for firewood, and villagers building a house must apply to them for roof beams and structural timbers. The shingi-nawa describe exactly where trees can be cut, usually not more than one in a given location.

Thirty-five years ago, Khumbu had virtually no schools, medical care, or basic amenities such as drinking water systems and improved bridges. Sir Edmund Hillary, who has devoted his life to helping the Sherpas of Solu-Khumbu District, knew that aid in the form of cash donations would do little to help. Since 1960, his Himalayan Trust has raised money

for a large number of projects, all in response to requests by the Sherpas. From their experience as traders—for which math skills and the ability to draft business agreements are essential—Sherpa elders have long recognized the need for literacy.

"Our children have eyes, but still they are blind," one elder said years ago. This general sentiment prompted the construction of several primary schools and a high school in Khumbu villages.

When smallpox was ravaging the Indian sub-continent, the Himalayan Trust was able to immunize most of the people of Solu-Khumbu. Virtually all new cases of goiter and cretinism were eliminated as well, following an iodization campaign that accompanied the 1966 opening of the Khunde hospital. Still, illnesses are often blamed on the influence of ghosts, and many Sherpas consult shamans before going to the hospital—a place where wounds can be treated, but where people have been known to die.

ONWARD

Each day of the approach, David and Steve prepared story boards and planned the day's film shooting, and David looked for images that would best capture Khumbu's 360-degree magnificence. Beyond Tengboche, they stopped on the Pangboche bridge, a suspended walkway 100 feet above the Imja Khola river. From here they could see the double hump of Ama Dablam (mother's amulet box), and the distinctive hanging glacier below the summit.

During the test filming a year earlier, David

asked that he be lowered by a rope into the gorge with the IMAX camera. "The shots don't come to you," he said by way of explanation. "I envisioned filming while being suspended in space, a helicopter shot done with ropes."

Fifty feet downriver from the bridge, he located a good anchor on one side of the gorge. From a tree on the other side, the Sherpas threw across a rope, and David tied into it. They then lowered him using a system of belay devices, while lowering the camera alongside. Once in position, David took the camera off the belay, clipped it on to his belt, and sent the rope back up for the batteries and lenses.

Across the gorge, more Sherpas hauled on the slack rope, creating a "V" with David in the middle. "To get me at the right height," David recalled, "it needed to be a wide "V", and took some tricky adjusting. If the rope was mismanaged, I would have swung like a pendulum and been smashed into the side of the gorge."

Yak drivers who were rounding the bend stopped when they reached the bridge, to let the yaks cross first. Yaks will spook if a load becomes snagged on the handrail or wire mesh of the bridge, and in their terror have been known to injure themselves, and damage the bridge and their loads. David was especially concerned about the camera gear, having already chased a recalcitrant yak that had turned around and trotted down-valley with some IMAX camera equipment.

Jamling saw the gorge as a good place for prayer flags: The strong wind there would carry the prayers aloft. He unraveled a string of flags and strung it along the rail of the bridge. He had already left prayer flags at two other power spots

along the approach route: above Namche Bazar and near the Dewoche nunnery. He would save the rest for the mountain, he decided.

Through the fog, the team hiked past simple stone houses, wind-stunted juniper, and more piles of mani stones. Beyond Pangboche, the vegetation gave way to rocks and lichens. Crossing a low pass within the valley, the Sherpas each selected rocks from beside the trail and placed them on a large cairn of stones, in thanks for the merit they would gain as pilgrims and for the good fortune of subsequent travelers.

THE ROUTE TO BASE CAMP

On March 28, the expedition took a detour to the 14,300-foot yak pastures of Dingboche. For several days they camped next to herders' huts that had been converted into solar-powered lodges featuring yak steaks, cinnamon rolls, and hot showers.

They filmed, washed clothes, and hiked to higher elevations to begin their acclimatization, the physiological process whereby the body's oxygen delivery system adjusts to the lower quantity of oxygen available in each breath. The more time spent at 14,000 feet, the better the team would be able to handle the two-day ascent to Base Camp, at 17,600 feet.

Ed, Jamling, and Jangbu had their hands full. Each load for Base Camp weighed about 66 pounds; yaks can carry two of these, but organizing a hundred yaks was a tall order. The cargo would have to be shuttled.

"It's not easy keeping track of 200 loads," Ed volunteered. "We haven't lost one yet, though at times they've been spread out over six

TOURISM BRINGS DOLLARS AND CHANGE TO THE KHUMBU

In 1964, the year the airstrip at Lukla was built, 20 tourists visited the Khumbu. At the time, Sir Edmund Hillary and anthropologist Jim Fisher had no idea Lukla would become a gateway into the Khumbu for thousands of trekkers annually. In 1995, more than 15,000 foreigners entered the region. Mingma Norbu Sherpa, Nepal's World Wildlife Fund representative, estimates that one member from every Khumbu household is now involved in the trekking industry.

Jim Fisher believes that modern tourism is akin to the ancient tradition of pilgrimage, which for centuries has been an organized and commercial undertaking. Indeed, many tourists proudly identify themselves as pilgrims.

"Truth to tell," Fisher said, "Sherpas are mystified that Westerners spend so much time and money to see what to them are sometimes sacred but not very interesting mountains. Even the most experienced sirdars admit they cannot fathom why foreigners climb, although they have hunches about motives—principal among them fame, money and science."

Dr. David Shlim has noted changes even during the twenty-four years the Himalayan Rescue Association Aid Post at Pheriche has been operating. "Recently, my wife and four-year-old son, and I

Proprietress of a trekking lodge—a Nepalese version of diner and motel combined—prepares tea. Dozens of such establishments throughout Khumbu offer travelers lodging and meals, which can range from yak steak to cinnamon rolls to pasta dishes.

sat at some outdoor tables at the edge of Khumjung, Ama Dablam massively at our backs, sipping cappuccino and eating fresh-baked pizzas from a large electric oven. It was different from sharing boiled potatoes around the

hearth of a Sherpa home, but it was hard to say it was less pleasant." Shlim feels that a bigger long-term threat comes not from tourists visiting Khumbu, but from Sherpas living most of their lives in Kathmandu. Many

Sherpa children don't learn the Sherpa language and can't imagine living full-time in Khumbu. "But it certainly is not our right to say that Sherpas can only be mountain guides and grow potatoes," he added.

"The new breed of Sherpas no longer wear sheepskin pants," Jim Fisher said, "but they know who they are. True, they wear down jackets, drink sugar (instead of salt) tea, and partition their homes into smaller rooms that are easier to heat, but these are superficial matters in themselves. What is more important is that Sherpas are proud of being Sherpas and Buddhists."

Climber Mike Thompson has also followed the Sherpas' fortunes and plights in the context of their broadening, changing world. "The new Sherpa entrepreneurs often come from less important families, and the well-established trading families have to some extent missed out. It is those who, having little to lose, were prepared to take big risks in the mountains that are now building hotels and shops in Namche and Kathmandu."

Many Sherpas have invested earnings from tourism in their religious institutions, and Jim Fisher believes that education, especially, has given Sherpas the tools to maintain their cultural equilibrium and allowed them to exploit the forces of change for their own benefit.

Hillary's Himalayan Trust established Khumbu's only high school, in Khumjung, and it sends qualified Sherpas for study abroad. Sherpa doctors trained in the West are now stationed at the Kunde and Phaplu hospitals, and others have been trained as park managers in New Zealand. One young Sherpa, Ang Zangbu of Jorsale, carried loads barefoot for years in order to pay for books and lodging at the Khumjung school. He became a jet airliner pilot at the age of 27, ten years after he set eyes on a motor vehicle for the first time.

Some have noted that Sherpas with political intentions have had to compete with and ingratiate themselves with Nepal's Hindu majority. In doing so, they have become masters at practicing what has been described as Nepal's three major religions: Hinduism, Buddhism, and Tourism.

Catering mainly to trekkers and tourists, shops like this one in Namche Bazar offer local woolens, carpets, and other handicrafts—as well as expedition leftovers such as propane stoves, crampons, dried and canned food, batteries, and paperback novels.

villages." Ed consulted his list frequently, which seemed redundant considering his near perfect recall of the contents and location of each load.

Ed and Jamling went ahead to assess the condition of the trail. Yaks attempting to reach Base Camp had been post holing and bellying out in drifts of snow, especially on north-facing slopes. The Sherpas refused to lead the yaks beyond Lobuche, a day's walk below Base Camp. They can be tempted by hazard pay themselves, but won't risk their yaks for any price. Ed and Jamling radioed to Jangbu to begin looking for porters.

The season's high snowfall, Ed felt, might at least make for better conditions on the mountain. Some of the rock would be covered with snow, making the climbing less technical. But the team remembered the tragic avalanches that killed over 60 people the previous November, after a cyclone in the Bay of Bengal veered north to the Himalaya. Trekking guide Brian Weirum reported that the nearby valley of Gokyo, a picturesque trekkers' destination known for its turquoise lakes and staggering views, received over ten feet of snowfall in 36 hours. There, a wet snow avalanche buried a yak herder's teahouse, killing 26 people, including 13 trekkers and their Sherpa guides and kitchen crew. Across the width of the country, more than 500 stranded people were rescued by helicopter.

A SACRED SITE

En route to Lobuche, above the hamlet of Duglha, the team rested at a place called Chukpö Laré, the site of a number of large stone cairns erected by the Sherpas. Chukpö Laré is referred to as a memorial site, but it is primarily a place of ceremony and ritual. If a Sherpa dies at Base Camp or on the mountain and his body is recovered, it is brought to Chukpö Laré for cremation.

Most of the expedition Sherpas have relatives who were cremated or honored here, and they all stopped to recite a prayer for their benefit. Jamling visited each of the more than 30 *chö-lung* monuments, and prayed and chanted. He stayed longer than the others.

"While returning from the summit in 1992, my cousin Lobsang Tsering Bhotia fell to his death. We're unsure of what happened, but he may have run out of oxygen and become delirious. Team members carried his body down here and cremated him."

Corpses are not empty vessels, at least not immediately, and a form of the deceased's spirit may continue to reside within the body, Sherpas believe. Thus it is best for one's reincarnation if lamas can be present to perform the proper rituals. The bodies left on the mountain create a special problem: Without funerary rites, malevolent spirits can linger in their vicinity and cause harm. Devout Buddhists, however, say that people with pure motivation will be little troubled by these wandering souls.

If possible, lamas officiate at the cremation, and they treat the fleshly body as a sacred offering; it is first purified, then given by fire. In a similar ceremony on the north side of Everest, dead Tibetans are flayed and left for consumption by vultures in what is known as "sky burial."

In the deceased Sherpa's home village, wealthy families will light 100,000 butter lamps in their private chapels and in the monastery, in an appeal for a favorable reincarnation. Monks are called to the Sherpa's home, too, and the soul is prepared for travel through the transitory, after-death state of *bardo.* On the 49th day after death, the person is reincarnated.

The ashes are then molded into clay votive tablets called *tsa-war,* which are returned to the memorial site and placed inside the *chö-lung* monuments. Sacred objects are enclosed with them, including a piece of juniper with carved inscriptions that must be oriented in the precise alignment it had as part of the standing tree, which is labeled before felling.

Ultimately, the chö-lung represent aspirations for a permanent state of peacefulness—of nirvana. Prayers recited here are said for all sentient beings. "In death, humans lose their individuality," Jamling explained. "That's why we discourage keeping souvenirs or remembrances of the dead, and names are not inscribed here."

Chukpö Laré is not an ancient site. In 1970, after six Sherpas were killed by an avalanche during a large Japanese expedition, the Incarnate Lama of Tengboche hiked up and sanctified the area. It is distinguished from other funeral sites in that the Sherpas cremated here all died untimely deaths, from unnatural causes. This complicates reincarnation and is difficult to resolve ritually.

Audrey Salkeld stresses that Sherpas have paid a disproportionately high price in lives lost on Everest. In 1922, seven Sherpa porters were buried under an avalanche below Everest's North Col, and in 1974 five were taken by an avalanche in the Western Cwm. In the first 70 years of Everest climbing, 54 Nepalese and Indian Sherpas were killed—more than a third of the total climbing deaths in that period.

Because of their contribution to route fixing and ferrying supplies, especially in the Khumbu Icefall, Sherpas are exposed to riskier parts of the mountain than their employers. Now, the sirdar's agency is required to carry a $4,000 life insurance policy for every Sherpa who enters the Icefall.

PORTERS TO BASE CAMP

When the yaks arrived at Lobuche, at 16,168 feet, the Sherpas unloaded them and joined the team members in sorting boxes and duffels for portering. Everyone was coughing and red-faced; even the veteran climbers weren't accustomed to the cold and altitude.

Porters were hard to find. Above the last villages it is difficult to recruit them in the best of conditions, and the high demand from all the expeditions had quickly led to astronomical pricing. But no challenge was too big for Jangbu. His perpetual grin, as deferential as it was devilish, betrayed a keen business sense. He had been issued a bright yellow walkie-talkie, and made sure he talked into it importantly when passing through villages and around young Sherpa women. He wasn't going to let the loads be delayed.

Before long, Jangbu was checking off names and load numbers on a rolled-up school notebook. "If Jangbu were American," Steve Judson said in amazement, "you'd find him living in southern California, running a successful shipping company, and spending most of his time surfing."

To leverage their incomes, each porter staked claims to three or four double loads, at

140 pounds each. Suspending them from their heads in simple handwoven tumplines, they ferried the loads in stages. One load went partway, then they returned for a second load, which they dumped with the first, and did the same with a third. Then they moved on to shuttle the next leg. In this manner, while deadheading downhill they could rest up for the next short-stage load carry.

From Lobuche, Base Camp was no easy jaunt. Many trekkers never reach it, and Steve briefly suspected that his light-headedness from hypoxia—the shortfall in oxygen resulting from reduced atmospheric pressure at higher altitudes—might keep him from getting there. "The trail seemed to change form," he said, "and my feet became disembodied appendages that found each step as if on their own." Robert stopped and threw Steve's pack on top of his, and watched him for any signs of his hypoxia progressing into altitude sickness.

As usual, Araceli sang to herself as she walked with the rhythmic, buoyant energy of someone who might break into dance at any moment.

On foot above the clouds, trekkers pass a string of cairns erected by relatives of deceased climbing Sherpas, some of whom were cremated here.

FOLLOWING PAGES:
Amid spectacular steeps of upper Khumbu, yaks never break their stride. Deep snows near Lobuche village so hampered the animals that Sherpa porters had to haul all Everest *team gear the remaining seven miles to Base Camp— on their own backs.*

CHAPTER THREE

AT HOME ON THE GLACIER

"Were it 1,000 feet lower it would have been climbed in 1924.
Were it 1,000 feet higher it would have been an engineering problem."
—1938 EVERESTER PETER LLOYD, LOOKING BACK IN 1984

"Changba, the head cook, stood waiting outside the kitchen tent—a tray filled with cups of tea in one hand and a gallant welcome wave in the other," reported Liesl Clark, describing the team's April 3 arrival at Base Camp. "The terrain here is otherworldly, with ice spires and shiny blue pinnacles protruding from the rock-covered snow and ice. The rocks—which Roger Bilham informs us are a combination of granite and migmatite—look clean and regular, like Hollywood props jumbled into chaotic piles." ¶ Two weeks on the trail were over. The Sherpa support staff had arrived at Base Camp a few days ahead of the others, and had carved out tent sites from the clutter of ice and rock on the margin of the glacier. Much additional work would be needed to prepare their home for the next two months. A winding line of porters

Aptly named Kangtega—it means "horse saddle" in Sherpa—glories in full sun while trekkers on trail are locked in shadow.

99

from ethnic groups with names like Rai, Limbu, and Tamang worked its way toward camp. As the loads came in, the Sherpas checked off their numbers. Jamling directed them onward: sacks of grain, fresh vegetables, and cans of kerosene to the kitchen tent; other food and climbing hardware to the storage tent (referred to as the "7-Eleven"); film equipment to the camera tent; the satellite phone, computers, and printers to the "communications corner" of the dining tent.

"If you can break it down small enough to carry it, you'll probably find it here," summarized production manager Liz Cohen. She had her hands full keeping track of the finances, media, shot lists, and communication with the outside world.

Both the climbing and film teams were unacclimatized to the 17,600-foot elevation. "Just existing is an effort," Roger explained. "While sitting, you feel relatively normal…but then stand up…and you can't form a sentence…of more than four words or so…because you're breathing so hard."

Beyond camp was the dreaded Icefall—the Khumbu Glacier's tongue of unstable ice that rudely protrudes from the valley between Nuptse and Everest. David and the climbers stood eyeing the treacherous Icefall, wondering quietly about the condition of this year's route through it.

The amphitheater of mountains surrounding Base Camp echoed the sounds of rumbling, billowing avalanches. Huge chunks of ice would tear off the Lho La, the wall that leads to the base of Everest's West Ridge, and come crashing down with thunderous roars.

"Lying in our tents at night, it sounded like the *1812 Overture*," Audrey Salkeld recalled.

The sound and motion swelled beneath them, too; Base Camp is situated on the edge of the glacier itself. "The Khumbu glacier is ice under stress: creaking and bumping and clicking and snapping and cracking and squeaking—a constant babble," Roger said. "It's always reminding us that we're camping on a dynamic sheet of ice."

Just as dynamic are the flocks of yellow-billed choughs, soaring on updrafts, their movements matching the playfulness of the wind itself. They've been seen poking about the South Col, at 26,000 feet, along with another Base Camp denizen, *gorak*—ravens—which have cadged food even higher. Gorak, the Sherpas say, can be messengers of the deceased and bearers of their souls.

HIGH-ALTITUDE CUISINE

Each morning, Changba woke before sunrise to prepare the first of endless rounds of tea, which he delivered to the tents amid clouds of steam and hearty good mornings. All the while, he carried on in cheerful song, the lyrics alive with the joy and angst of mountain life.

Audrey found her spot in the kitchen, and filed her second report to MacGillivray Freeman Films and the NOVA Web site:

"Base Camp is now quite a city.…It is curious that, having voluntarily removed ourselves as far as possible from the trappings of so-called civilization, expeditions then appear to vie with one another in creating alternative civilizations of ingenious comfort and complexity. Elaborate mess tents have sprung up, with electric lighting and in some cases heating, music, and comfy chairs and tables. Even the Sherpas compete in the construction of impressive camp kitchens, mostly sangars (dry stone walls) with pitched roofs made from heavy-duty tarpaulins and laid out inside with all the economy and efficiency of kitchens in the best hotels."

The kerosene and propane stoves were always going, and the climbers and film team hung out in the kitchen for warmth. Sherpas visited from other camps, and the Sherpa women who came up with loads laughed and joked constantly. The Sherpas invited the porters in and fed them—then put them to work out back, doing the dishes.

The shelves on one wall of the kitchen tent were weighted with cans and boxes, and a leg of yak hung from the crossbeam. Araceli's mother had sent her off with a large ham, as she had for each of her daughter's Himalayan expeditions.

"I brought something very tasty, too," Robert said in his Austrian German accent. He waited for Araceli to respond.

"*Zumzing?* So what does zumzing taste like?" Araceli teased, her attempt at a German accent enhanced with a Catalan inflection. Robert showed her his large wrapped *speck,* a smoked ham from Austria "loaded with valuable calories." To wash it down, he had packed alongside it a liter of Voglbeer schnapps. Robert proudly told her that the voglbeer bush, which grows at tree line near his home, supplies the fruit for this distillate.

"I want to see you after you have drank all of that and are destroy-ed," Araceli gibed.

Jamling had brought his own specialty: four kilograms of *tsampa,* which he would take to the higher camps. This roasted barley flour digests easily and provides lasting fuel, though a handful of dry flour can be hard for the uninitiated to get down. Some Sherpas carry a small leather sack in which they mix tsampa, tea, butter, and sugar, and knead them into dough balls called *pak.* Nuts and other ingredients are sometimes added to form a Sherpa version of a high-energy bar.

On April 10, Paula, Sumiyo, and Changba prepared a full Japanese dinner, but most meals were not so lavish, consisting of creatively disguised cooked potatoes.

One afternoon when Ed returned from a load carry to Camp I, he was sorting loads for Camps I and II. "How about Spam tonight, and maybe some wine?" he queried toward the kitchen tent.

"Spam?" Sumiyo said quizzically. Ed nodded enthusiastically to her, then handed a roll of duct tape to a climbing Sherpa who was repairing his snow gaiters.

"Spam," Sumiyo repeated. Pondering intently, she enunciated aloud to herself the other English words she had been practicing. "Revision… Tremendous… Awesome… Spam."

Araceli perked up. "Did someone say 'Spam?' I'm ongry."

"Are you *angry* that we have Spam, or are you *hungry* to have some?" Ed asked her, provoking boisterous laughter from the crew in the kitchen.

"No, *ongry,*" Araceli repeated indignantly, unfazed by the laughter. "I want some food, of course, like I do always."

"I think you get *angry* when you get *hungry,*" Ed joked. Araceli flashed a brief smile, then swatted him. A party of rose finches abandoned their foraging for scraps and fluttered off.

MOUNT EVEREST
29,028 ft   8,848 m

South Summit
28,710 ft
8,751 m

Dashed section of route
located behind ridge

Lhotse
27,890 ft
8,501 m

Northeast Ridge

South Col
26,000 ft
7,925 m

Camp IV
26,000 ft
7,925 m

Geneva

Spur

Nuptse
25,790 ft
7,861 m

Bei Peak
24,878 ft
7,583 m

Camp III
24,000 ft
7,315 m

West Ridge

Khumbutse
21,867 ft
6,665 m

Western Cwm

Camp II
Advance Base Camp
21,300 ft
6,492 m

*Rongbuk
Glacier*

Lho La

Lingtren
22,142 ft
6,749 m

Camp I
19,500 ft
5,943 m

Pumori
23,507 ft
7,165 m

Khumbu Glacier

Khumbu

Icefall

Base Camp
17,600 ft
5,364 m

Khumbu Glacier

1996 *Everest* Expedition,
South Col Route

Cho Polu
21,965 ft
6,695 m

Baruntse
23,517 ft
7,168 m

Namche Bazar
18 miles

BERANN

MORE THAN ONE WAY
TO CLIMB A MOUNTAIN

The 1963 American expedition to Everest put 6 of its nearly 1,000 men on the 29,028-foot summit, one pair laboring up the unconquered West Ridge while the others took the South Col route (red) pioneered by Hillary and Tenzing a decade earlier. Sherpas and climbers leapfrogged supplies up the forbidding slopes through a series of camps placed a day's climb apart. The 1996 ascent by the Everest *team also headed up the South Col, but pitched one less campsite, electing to "sprint" from Camp IV to the top and back in an 18-hour marathon climb.*

Each night Jamling would check out what the Westerners had cooked and then what the Sherpas had prepared, before choosing where to eat. A sip of wine would be hard to pass up, he decided.

Evening dinner dress consisted of a down parka, hat, and gloves—donned each afternoon during "down up time," the moment the sun went over the ridge. Base Camp visitors noticed that the dining room was as cold as a meat locker, but hunger made the food delicious. After dinner, team members stumbled through the dark to their tents. Despite the cold, porters stood around in the evenings in rubber flip-flop sandals without socks, swearing that they felt fine.

"We're on our honeymoon here," Ed said. "Living dormitory style we don't really have a chance for privacy, but I'd rather have Paula on this trip than write letters to her."

GLACIER GADGETS

Within a week the team had regained some strength. "Since we arrived," Roger reported cheerfully, "I've been mending the electronic and solar gadgets that inevitably break, while working in tiny, cold corners with limited resources—I love it!" Already he had fashioned a voltage transformer from parts he had found in the Asan Tol bazaar.

It took filmmaker Brad Ohlund a little more time to adjust to glacier life. "I live on the beach in southern California, and most of my favorite places have jungles and warm water," he

White thunder: One of endless avalanches booms and billows down from the Lho La within sight of Base Camp. Avalanches are the single biggest cause of death on Mount Everest.

explained. But he had developed a symbiotic rapport with Steve, Robert, and David, each of them dedicated to getting a properly functioning camera to the summit—and back safely. The preparation required long hours from Brad, who was responsible for overseeing the complex camera gear—film and magazines, filters, matte boxes, battery packs, shades, tripods, countless accessories, and the unwieldy camera body itself.

The fax machine and satellite phone drew more juice than the jury-rigged solar system could provide, and the generator was frequently needed as a backup. In the minus 16° F mornings, starting it required a hundred pulls on the cord and a magical blend of adjustments to the choke and throttle.

The "sat" phone did get some use. On April 15, Dr. Charles Warren, one of the few surviving pre-war Everesters, was delighted to receive a call from Audrey and the team, wishing him a happy 90th birthday. A member of the '35, '36, and '38 expeditions, Warren was the first to discover the body of Maurice Wilson, the eccentric pilot who in 1934 hoped to reach the summit through fasting and prayer. Warren and his companions committed the remains of "the Mad Yorkshireman," as the press dubbed him, to a crevasse below the ice chimney of the North Col. They then sat beneath an overhang and opened his diary, which made for eerie reading.

The chat with Warren launched Audrey into after-dinner stories, and she spoke of the film that she

and David had scripted and produced on the fate of George Mallory and Andrew Irvine, the two British climbers who disappeared high on Everest in 1924. In her blue Volkswagen bug, David and Audrey crisscrossed England seeking out other veterans of early Everest attempts. They first visited Capt. John Noel, photographer and filmmaker for the 1922 and 1924 expeditions, in his weatherboard cottage on Romney Marsh in Kent. Noel had been an artillery instructor during the Great War, and he advised David to launch a succession of small "eggshell" bombs high on Everest to excavate the ice and snow, in the hope of finding traces of the two men. Then they motored to Cambridge to interview another nonagenarian, Professor Noel Odell, the last person to see Mallory and Irvine alive. In Bakewell, Sir Jack Longland described how, in 1933, his team had discovered an ice ax high on the mountain,

"KNOCKING THE BASTARD OFF"

BY AUDREY SALKELD

When Sherpa Tenzing Norgay came to Everest with the British in 1953, it was his seventh expedition to the mountain. The previous spring he had climbed to within 820 vertical feet of the summit with a Swiss party and liked to joke afterward that had he and Raymond Lambert only been able to brew a cup of tea, they would have made it to the top.

Unlike most Sherpas, for whom working with expeditions is merely one of few job opportunities available, Tenzing was a mountaineer by choice and openly ambitious to stand on top of the world. So too was New Zealand beekeeper Edmund Hillary, who reflected in his memoirs that if you accepted a measure of ruthlessness and selfishness as necessary to success, then in 1953 he and Tenzing came the closest to the climbing prima donnas of modern times. Nobody worked harder than they to make the expedition successful, but in their minds success was always equated with them both being "somewhere around the summit" when it happened. It was no wonder that their leader Col. John Hunt quickly paired these two as a likely assault team.

All the early British expeditions to Everest had approached from the north, through Tibet, a route now closed to foreigners since the Chinese occupation of that country in 1951. Eric

Shipton's reconnaissance of the Khumbu Icefall in 1951 demonstrated that the peak might be climbed from its southern flanks in newly opened Nepal. Immediately, the Himalayan Committee in London began preparing a full-scale climbing attempt, and news that Swiss mountaineers had been granted sole access to Everest in 1952 came as a bitter blow. Polite overtures, aimed at making this an Anglo-Swiss venture, broke down over the question of leadership. It was hard to keep nationalism out of something as momentous as climbing Everest. The Swiss set off for the moun-tain—and the British bit their nails as they waited in line untill the following year.

At least the delay enabled men and equipment to be tested on nearby Cho Oyu. When their competitors failed so narrowly, British relief was tempered by a sense of urgency. There would be no second chances. The mountain was already booked ahead by other mountaineers. John Hunt was recruited to replace veteran Everester Eric Shipton, who it was popularly assumed would lead the expedition. This was a decision taken after awkward backdoor maneuverings on the part of the committee, but even

those who saw it as a cruel snub to the man who had done so much to pioneer the way were eventually won over by Hunt's dedication and fairness. Shipton himself declined a place in the party, having little sympathy for the type of expedition this was turning out to be.

Hunt threw himself whole-heartedly into building a happy and cohesive team. To get climbers and their tents, food, and oxygen in position for three summit attempts, tons of bag-gage and many men would have to be deployed for a long siege in one vast pyramid of effort. Hunt knew they must move speedily at

Exhausted but triumphant, Tenzing and Hillary relax at the expedition's Camp IV, some-where around 24,000 feet, after their historic first ascent of Everest on May 29, 1953.

altitude to cover ground before deterioration set in, or be so fortified as to slow down the deterioration. It pointed to the use of bottled oxygen, which had always been an unresolved question of conscience for earlier expeditions. This time, the moral decision was taken before anybody left home: All who went high would use supplementary oxygen.

The party set out on foot in the second week of March from Bhadgaon, near Kathmandu, for the 17-day trek to Tengboche, below Everest. Two further weeks were devoted to acclimatization, route-preparation, training, and general shaking-down as a team before the first climbers reached Base Camp at 17,900 feet on the Khumbu glacier. Staging camps were established in and beyond the Icefall. Their Camp IV at 21,200 feet in the Western Cwm became Advanced Base, and Camp V was set 800 feet higher at the foot of the Lhotse Face. From here the team would climb the face, as the Swiss had done the previous autumn, before traversing the top of the Geneva Spur and dropping down to the South Col. Camps VI and VII were duly installed on the steep face, and Camp VIII, on the Col itself, was reached on May 21 by Wilfrid Noyce and Sherpa Annullu.

Above this desolate wind funnel of blue ice and rubble

rises the South Summit of Everest, which has to be surmounted before you reach the main summit. It had not been trodden before: Lambert and Tenzing were forced back just short of its crest.

On May 26, Hunt and Da Namgyal struggled upward with loads for an assault camp, to be placed as high as possible in preparation for an all-out attempt by Hillary and Tenzing. Meanwhile, Bourdillon and Evans were already making the first summit bid even though, by starting out from the South Col, they had little chance of going all the way. Their principle task was to open the route over the South Summit.

Evans' oxygen apparatus gave him trouble most of the day, and the route proved trickier than expected. At 1 p.m. the two men scrambled on to the South Summit to see the final, narrow, corniced ridge switchbacking ahead, with alarming drops on either side. They had climbed higher than any man before, but could not safely go farther. When they finally made it back to the South Col, utterly exhausted, their faces hung with frost, they looked like strangers from another planet.

Two days later, Hillary and Tenzing moved up to occupy the small assault Camp IX at 27,900 feet. Tucked on a tiny split-level platform overhanging the tremendous South Face, they passed a

wild night, gusty and cold. Sometime around 4 a.m. Hillary looked out to see the makings of a perfect day. Stomping his feet to get warm, Tenzing let out a whoop of delight when he spotted the Tengboche Monastery miles below in the blue shadows of the valley. It seemed a good omen. They drank lemon juice and sugar, nibbled biscuits, and Hillary thawed out frozen boots over the primus stove. At 6:30, wearing all the clothes they possessed, they crawled outside. This would be the day—May 29, 1953—which would change the lives of everyone on the expedition.

Though still dark on their ledge, the way ahead was bathed in sunlight, beckoning them upward.

"Let's go!" Hillary urged and, grinning, his companion scrambled past him to kick a long line of steps back up to the main ridge.

The snow crust demanded care. Hillary took over the lead and painstakingly packed down the snow in each foothold as he forged over the South Summit. Conditions improved. The two picked their way delicately along the virgin ridge, between the cornices overhanging the east face and the abrupt drop of the southwest face. After an hour or more of steady going they came to a steep, rocky step, some 40 feet high. Its existence was known in advance from aerial photographs, but no one knew

whether or not it could be climbed. Luckily, Hillary was able to wedge himself half into a crack and wriggle upwards. The obstacle is still called the Hillary Step today.

Tenzing followed up, and the pair continued their switchback progress along the summit ridge until, having passed the last corner, ahead lay only a snowy dome and the vast plateau of Tibet. "A few more whacks of the ice ax in the firm snow, and we stood on top," Hillary said later, unwilling to spell out that he had made it a few paces ahead of his partner. He sought to shake hands, but a delighted Tenzing would have nothing but to fling his arms around his friend's shoulders and thump him vigorously on the back. It was 11:30 a.m. and the highest pinnacle of the world was at last trodden by man.

The Everest achievement, inextricably linked in history with the coronation of Queen Elizabeth II, was seen within the British Commonwealth as heralding a new and glorious era. Nowadays, we are more likely to regard it as the last imperial adventure. Either way, there was nothing foregone about the conclusion. Every success requires an element of blessing, but above all this was the result of a magnificent team effort grafted onto the hard experience earned over more than 30 years of Everest expeditions.

which could only have belonged to one of the missing men.

Audrey and others speculate that if the remains of Mallory and Irvine are found, their camera might contain exposed film that—preserved by the cold—could be developed. If they had reached the summit, it is inconceivable that there would not be a summit shot to prove it.

Ed spoke unequivocally. "Even if Mallory and Irvine touched the summit, they didn't *make* it—that's like swimming to the middle of the ocean," he asserted. Sir Edmund Hillary is of the same opinion. "The point of climbing Everest," he says, "should not be just to reach the summit. I'm rather inclined to think that maybe it's quite important, the getting down."

CROWDING IN THE MIDDLE OF NOWHERE

By mid-April, Everest Base Camp had filled with 14 expeditions, and their camps stretched for a mile along the rumpled edge of the glacier. The climbers, Sherpas, Base Camp staff and government liaison officers brought the Base Camp population to more than 300.

Araceli, accustomed to climbing in small parties on remote routes, was feeling crowded. "The approach to here has become commercial—you can even find sneakers to eat, just two hours from Base Camp!...I mean *Snickers,* the chocolate, not the shoes!" She laughed and drew a line around her mouth to show where the chocolate usually becomes smeared. "Oh, and you can find the shoes, too."

Many of the leaders of these

Ever present prayer flags crown one corner of the ephemeral tent city known as Base Camp, just downslope from Khumbu Icefall. The arrival of the Everest *film team and 13 other Everest expeditions in 1996 caused the Base Camp population to swell beyond 300.*

expeditions—David Breashears, Scott Fischer, Rob Hall, Todd Burleson, Pete Athans, and Henry Todd—were eminences in the world of guided climbing, and all were friends. Nonetheless, Base Camp hummed with an undercurrent of competition among the climbers, and between the groups. To make a living climbing, one must be known and recognized. This recognition is achieved mainly by assembling a portfolio of successfully climbed peaks. It is to the guide's advantage, professionally, to excel where others haven't. Rivalry could be expected.

This year, Scott Fischer's and Rob Hall's expeditions were the largest. Fischer's Mountain Madness camp displayed banners advertising Starbucks coffee, and Sherpas painted "New Zealand camp" in large letters on a glacial boulder at Hall's campsite.

Scott Fischer moved about Base Camp informally, and his warmth and positive approach infected everyone he met. Ed had climbed K2 with Fischer in '92, and Scott had photographed Ed and Paula's wedding, only a few weeks earlier. "Scott is freewheeling and almost casual about organization," Ed said. "He assumes that the details—how much fuel or how many ropes are needed—will fall into place, and for him they usually do. 'Forge ahead! We're going for it!' is his motto. He's fun and inspiring, overflowing with energy and enthusiasm. But sometimes I feel he could use some reining in."

Ed described Hall as the exact

opposite: calculating, meticulous, and organized. "Rob is always assessing conditions, the
weather, what the clients are doing, asking
himself '*what if?*' He uses his incredible drive
to help people achieve their dreams, and takes
joy in a task well done."

Ed respected Hall for his climbing expertise
and experience, and Scott Fischer was planning to join the two of them on a climb of

Manaslu immediately following Everest. Ed
had guided several commercial climbs for
Hall's New Zealand company, Adventure
Consultants, and he anticipated a long-lasting
partnership.

Climbers and clients this year had come to
Everest with a variety of motives, not all of
them clear.

"Egos and other factors are at work here,"

Paula said, voicing a concern that Ed had shared earlier. "People can become selfish on the mountain, and lose sight of what's important." Veteran climbers worry that many clients regard an attempt on Everest as a once-in-a-lifetime chance—predisposing them to push their luck and take risks, without the experience that would help them gauge those risks. Out of seven Everest expeditions, Ed has summited three times. Most other experienced guides have even lower summit-to-attempt ratios. Nonetheless, during the approach to the mountain, Scott Fischer expressed a modern and not uncommon sentiment about Everest: "I mean, honestly speaking, we have learned how to climb Mount Everest, and are building the yellow brick road to the top."

It was a road that didn't require a driver's license. The Taiwanese team in particular was underexperienced, though it would be hard to fault them for trying. Audrey noted that several of the Taiwanese were missing fingers that had been lost to frostbite on previous climbs. "Imagine gaining your experience for Everest in a country as flat and warm as Taiwan," David said, adding that the

Mainstay of modern team expeditions, Sherpa porters head out from Base Camp heavily laden with gear and supplies needed to establish higher camps along the route. Climbers on guided expeditions generally follow only after they have acclimatized to Base Camp altitude—and tents have been pitched farther up. Historically, some groups set as many as nine successive camps; today, most rely on only four.

Taiwanese reminded him of Mallory and Irvine: sporting and eager, but poorly equipped and in over their heads.

The most unusual approach was being made by a lone Swede, Göran Kropp, 29. Remarkably, he had bicycled from Sweden to Nepal with most of his gear, and intended to solo Everest in the purest sense of soloing—he would accept no help, not even a cup of tea from others on the mountain. He was hauling his own supplies through the Icefall on a route he had found himself.

MINDFULNESS

Jamling stood quietly studying the sprawling, impromptu tent city. He spoke with Wongchu about Geshé Rimpoche's lesson to him on Right Motivation. Just that day, Wongchu had lined up a group of young, playfully rebellious Base Camp support Sherpas. He lectured that he would fine them 10,000 rupees each if they didn't cooperate with Changba, if they weren't helpful to the team, or if they were involved in any extra-marital sex at Base Camp. The Sherpas knew well that the last would offend the deities. Wongchu was especially concerned that any activity that generated emotions such as anger, jealousy, lust, or pride be avoided when on the mountain, for these would affect one's mindfulness when climbing.

"Which flag should we fly, were we to fly one?" Robert mused at dinner one night. Jamling pulled out the string of flags he intended to display on the summit: Nepal, India, Tibet, U.S.A. and the United Nations.

"My parents are from Tibet, but lived for long periods in Nepal and India, where I was raised. I studied and worked for years in the United States. The UN flag might represent me best—and our team as well. Look at us here." He nodded around the dining tent. "Most of us are from different countries."

He gently passed his hand over another multicolored bundle and carefully untied it. "The most meaningful flag for me is this *lung-ta,* the prayer flag. Lungta is actually a "wind horse" that bears a deity carrying wish-fulfilling gems, and their image is printed on many of the flags. But lungta also represents the degree of positive spiritual energy and awareness that propels a person—it's their level of divine inner support."

The Sherpas say that if their lungta is high, they will survive most any difficult situation, and if it is low, they can die even while sitting at home. A former Tengboche monk now working in the United States described the relationship between lungta and "good luck" as similar to that between principal and interest. "Cultivating one's lungta, through meditation and right actions, helps generate clear thinking," Jamling counseled.

He pulled out the other items he hoped to place on the summit: pictures of his mother and father, neatly framed in a red vinyl wallet; a photo of His Holiness the Dalai Lama; and an elephant-shaped rattle that his infant daughter selected from a pile of toys—perhaps significant in light of the Rongbuk lama's translation of "Chomolungma": Great Elephant Woman. Tenzing Norgay had also taken a small toy to the summit, selected for him by Jamling's sister.

"If we go for the top on the 9th of May, I

feel that my father's spirit will be with us. That's the tenth anniversary of his death."

"Jamling and I shared a tent," Roger remembered, "and one evening he lent me a signed edition of his father's book describing the first climb. I read it by flashlight as he recited his prayers. I asked him what he was praying for. He said, 'I pray for the safety of all, and that I am worthy to follow in my father's footsteps.' It brought tears to my eyes. I always looked forward to chatting with Jamling, because our conversations always went slightly beyond the present;

Joint of yak makes for a happy cook in the team's well-stocked Base Camp kitchen tent. Most expeditions spend about two months total time at Base; some arrange for periodic deliveries of fresh bread and vegetables, by yak train.

they were philosophical, never involving any shared experience."

PUJA AT BASE CAMP

Sherpas and climbers normally won't climb above Base Camp until the puja ceremony of purification and propitiation has been completed. Each expedition performs its own puja, and they begin by building a handsome chorten-like structure, about eight feet high, which becomes the heart of the *lhap-so,* the worship site. Wongchu selected an auspicious day from the Tibetan calendar, then

summoned an elderly monk from Pangboche.

The Base Camp puja is commonly described as a request for permission to climb the mountain, and for protection and good weather. But its liturgical meaning is much more complex. It is a category of *serkim* ("golden drink offering") ritual: before any new undertaking such as building a house or climbing a mountain, a lama engages the deities and asks for their understanding and toleration of their activities.

The morning of the puja, Sherpas and team members brought offerings—bread, rice, barley, and fancier items such as chocolate and whiskey—to the lhap-so. The lama sat, and two Sherpas poured him tea and made him comfortable. As he read prayers aloud, some of the congregatation meditated and prayed, while others moved about informally, as one may at most Sherpa ceremonies.

Sherpas believe that the gods will provide for the climbers—or deliver bad luck—even without this ceremony. But like meditation, the ritual is a form of discipline that opens one to receive the wisdom of these gods, empowering the faithful to better recognize beneficial or ominous situations when they arise. The serkim effectively activates the levels of concentration and awareness that are needed to succeed.

Not only must the climbers and Sherpas be purified before setting foot on the mountain, so must their equipment. In one part of the ritual, Sherpas ignite juniper boughs next to the altar. The team and climbing Sherpas then pass their ropes, crampons, ice axes, and other gear through the smoke, bathing it in protective wafts of incense. In the same way that the sweet smell of the juniper incense expels odors, the smoke dispels pollution and clears the way for favorable events.

The lama distributed red blessing strings called *sungdi* (from *sung-dü,* protection knot), made of thin braided nylon. The team and Sherpas tied one sungdi onto every item of equipment, to be left on for the duration of the expedition. "We Sherpas believe that sungdi—like the *sungwa* amulets—help protect us from the harmful spirits and situations that can push us into a crevasse, off the edge of a narrow ridge, or into the path of an avalanche," said Jamling. "But we recognize that these alone are insufficient to save us, of course."

Two Sherpas erected a tall prayer flag pole and secured it in a stone enclosure. Seven colorful strings of prayer flags radiated from the top of the pole, their ends anchored nearby. The Sherpas say that if a gorak lands on the juniper branch tied to the top of the pole during the puja, the expedition will be successful. If the pole breaks or is taken down, bad luck will ensue.

To close the ceremony, everyone sang in unison in a gradually rising tone, *Swöööööö!*—Go up, may good fortune arise!—while slowly

In an enduring Base Camp tradition, ice axes and other climbing gear await ritual blessing at a flag-bedecked lhap-so—*worship site—before team members proceed higher.*

FOLLOWING PAGES:

Tossing offerings of barley flour skyward, climbers and Sherpas at a Base Camp lhap-so *conclude the puja, a ceremony of propitiation and supplication.*

elevating right hands full of tsampa. Repeating this a third time , they launched the flour skyward. In a joyous, chaotic moment, all shouted "Lha Gyalo!"—May the Gods be Victorious!—and rubbed the flour in one another's hair and on their cheeks to signify that they wished to live until their hair and beards turned white.

The lama poured chang as a communion. Jamling accepted some in his right hand, his left held respectfully beneath it. He drank a sip, then ran the rest through his hair to fully incorporate the blessing. The climbers followed, and the offerings were passed out. "We all

Buddhist lama from a nearby village leads a Base Camp puja in prayer. Celebrants seek both blessing and permission from the mountain's spirits to proceed, often making offerings of food—which may range from barley flour to whiskey and candy bars.

accepted some of the offerings, careful not to take too much," David said.

"You must not argue on this day," Wongchu cautioned. They didn't look as if they would: Everyone had begun dancing and drinking—in moderation.

CHECKING THE OXYGEN

Geshé Rimpoche's lesson on right mindfulness called for attention to every safeguard. David inspected each of the team's 75 oxygen bottles and every regulator before sending them through the Icefall. The bottles were a lifeline, a ticket home. "We'll be very exposed above the South Col, where it's like being on the moon," he said, referring to the low atmospheric pressure—and likely to the hypoxia, which imparts the sensation of not being fully earthbound. Though David expressed unease at the team's dependence on this technological crutch, he handled the bottles with a respect bordering on the religious.

Most climbers prefer the Poisk high-pressure oxygen system. Each of the Russian-made bottles weighs 6.6 pounds when full and provides for six hours of oxygen flow at two liters per minute. The bottles are made of aluminum wrapped with strands of Kevlar, which help contain the 14,000 pounds of pressure. They are filled with pure oxygen, which mixes with outside air when the climber inhales.

Except for Ed, who will be attempting to climb Everest a third time without bottled oxygen, the team will use oxygen when climbing above Camp III, at 24,000 feet. At Camp IV they will sleep with it on at a flow of about a half liter per minute. On summit day they will each consume three bottles.

WHAT OXYGEN DOES

Dr. Charles Houston, a pioneer of modern high- altitude physiology studies and an adviser on the film, explains that when working muscles don't receive enough oxygen, they can "go anaerobic", and begin burning fuel without oxygen. But this results in acidosis, which the body doesn't tolerate for long. Breathing bottled oxygen increases arterial oxygen—and thereby hemoglobin saturation—which brings more oxygen to hungry tissues. And with more oxygen available, the extreme urge to breathe is

reduced, giving both the feeling and reality of more energy.

Oxygen was first used in the 1920s, but with criticism from high-mountain purists—even though it was not clear at the time that climbing Everest without it would be possible. Until 1946, that is, when Houston oversaw Operation Everest, in which four human subjects in a decompression chamber were taken slowly (over a period of one month) to the decreased atmospheric pressure of Everest's "summit." There, two of the subjects were able to perform light work for a short period.

Houston pointed out that if Everest, at 28 degrees north latitude, were located at the latitude of Denali, 63 degrees, it would have the barometric pressure of a peak at least another 500 feet higher—possibly making it unclimbable without bottled oxygen.

For emergencies, the team would have access to a novel item: an inflatable hyperbaric bag made of airtight, reinforced vinyl into which a very sick patient can be placed. It does the opposite of Houston's chamber: As the bag is continually pumped up by helpers, the air (and thus oxygen) pressure can be increased to that of lower altitudes. Afflicted climbers can improve their condition to the point where they are able to descend with minimal assistance.

ALTITUDE, SHERPAS, AND GEESE

Filmmaker Steve Judson sat on a rock one morning, a few days after arriving at Base Camp. "It's humbling, when I feel so bad, to watch the Sherpas scampering around as if they were at sea level." Aside from having had a longer opportunity for acclimatizing, many wonder if the Sherpas and other highland peoples are physiologically different from the rest of us. Physical anthropologist Dr. Cynthia Beall believes the Sherpas may possess a gene that allows more efficient oxygen delivery, giving them an advantage over lowlanders.

One part-time Himalayan resident, the bar-headed goose, is an even more remarkable acclimatizer in this regard than the Sherpas and Tibetans. High-altitude physiologist Dr. Robert "Brownie" Schoene is impressed that these birds can "winter in the marshes of India, then—without the advantage of gradual acclimatization—get up and fly over the Himalaya to their summer breeding areas high on the Tibetan Plateau."

Like all birds, bar-headed geese have a remarkably efficient one-way flow of air into their lungs, which eliminates wasted breathing. But they also possess a hemoglobin molecule in the red blood cells that distinguishes them genetically from their lowland bird relatives. "This hemoglobin has a greater capacity for picking up oxygen in conditions of low oxygen pressure, and is able to unload that oxygen to the tissues that are starved for it," Schoene explained.

Thomas Jukes, a biologist who has studied bar-headed geese, suggests that one day Himalayan climbers might be able to receive transfusions of a similar, genetically engineered hemoglobin before ascending to high altitudes.

CAMP-BY-CAMP SIEGE

Most teams place their series of four mountain camps at locations that have been used for the past 40 years—sites that were initially selected

for their relative safety from avalanche, rockfall, wind, and movement of the glacier. Each expedition groups its tents closely together, though these may be up to a few hundred yards from another expedition's camp.

After a route has been found through the Khumbu Icefall, the goal of every team is to establish and fully stock their four camps with supplies, so that the summit push can be made quickly from Base Camp or Camp II. This process generally takes a month or longer.

During the first week on the mountain, the Sherpas typically rise early, and in a couple of hours carry a 40- to 50-pound load to Camp I, at 19,500 feet, then return to Base Camp. When much of the gear for the higher camps has been carried to Camp I, the gear is shuttled to Camp II, at 21,300 feet. There, the Sherpas set up a smaller version of Base Camp—Advanced Base Camp, with a cook tent, dining tent, and individual tents. Most of the climbers' rest time above Base Camp is spent here. When acclimatized, they can climb directly from Base Camp to Camp II, and Camp I mainly becomes a way station for gear.

At Camp III, 24,000 feet, the climbers need only two tents; the Sherpas do not stay here, but generally go directly from II to IV.

Camp IV, the uppermost camp, is situated at 26,000 feet on the South Col, a broad saddle between Lhotse and Everest. The South Col is higher than all but 17 of the world's highest peaks, placing it in the "death zone," a poorly defined but easily recognized altitude where climbers know they must limit their time as their condition deteriorates fairly quickly above this altitude. Here, six tents are

erected, but until the night before the summit push, Camp IV is little more than a place to store oxygen and equipment.

The day before the summit attempt, the team climbs from Camp III to Camp IV, arriving there in the early afternoon. Here, they rest for a few hours and rehydrate themselves, or "brew up," before departing about midnight for the summit. They generally reach the "Balcony" at the base of the Southeast Ridge around sunrise, then continue along the ridge to the South Summit. From there, they negotiate a traverse to the the Hillary Step, a 40-foot-high wall of rock and ice, which,

Like a conquering army, an Everest expedition travels on its stomach. Members need to maintain strength and body weight as they adjust to oxygen deprivation; food and other supplies must be pyramided to successively higher camps in a repetitive but necessary operation that assures climbers their best shot at the top.

when surmounted, puts them within an hour of the summit. Hopefully, the climbers will return to Camp IV before dark, after 18 hours of climbing.

Camp IV is the Camp VI of the early 1950s, and Camp VII then was placed on the Balcony, at 27,600 feet. Only a few believed that climbers could reach the summit and return to the South Col in a day. Though physically just as difficult, the summit push may be psychologically easier, now. Climbers seem more willing to climb longer in exposed situations.

"Camp IV is needed, but it's basically a rest and rehydration stop," Ed noted. "You get there

WHAT HAPPENS AT ALTITUDE

BY CHARLES S. HOUSTON, M.D.

It's a law of nature: Without oxygen, we die. At high altitude, oxygen is in short supply, and simply getting a good breath is a challenge to any climber.

Oxygen density at varying altitudes is a function of another basic law: gravity. Air has weight. At sea level it is compressed by the blanket of air above it, and is comfortably dense. At high altitudes, air pressure is lower and air is consequently thinner. Although it contains the same 21 percent oxygen as it does at sea level, less oxygen is delivered per lungfull at higher altitudes. Adjusting to reduced oxygen levels requires some rapid physiological changes and is followed by slower, long-term changes.

The first change is obvious. We breathe harder to get more oxygen into our lungs. When we inhale, air rushes deep into the small air sacs of the lungs (the alveoli), where oxygen diffuses into the blood in adjacent capillaries. There, the oxygen bonds loosely with hemoglobin, the large iron-containing molecule in red blood cells, which carry the oxygen through ever smaller blood vessels to individual cells.

As oxygen passes through each stage of its journey from lungs to cells, it loses pressure in steps referred to as the "oxygen cascade." At oxygen's destination, the cells require only a small amount of oxygen to function; but if this pressure is too low, the continuous passage of water and electrolytes across cell membranes is disturbed—along with basic cellular functions.

Breathing harder provides the lungs with more oxygen, but also causes problems with loss of too much carbon dioxide. As a by-product of metabolism we need to exhale carbon dioxide, a swift and sensitive agent for adjusting the acidity of blood. Overbreathing "washes out" carbon dioxide, causing low acidity, or alkalosis—an abnormal condition that cannot be tolerated for long and that can cause light-headedness, fainting, spasms, and death.

As we breathe faster and deeper, the heart also beats faster and harder, pumping more blood per beat and per minute. The small capillaries in tissues become more permeable when short of oxygen, causing fluid to leak into the tissues and concentrating the red blood cells, thereby increasing the blood's oxygen content.

These changes—increased breathing, higher output from the heart, and the concentration of blood—are "struggle responses." They protect the body from oxygen deprivation, or hypoxia, but they can't be sustained for long without provoking potentially life-threatening symptoms. Acclimatization to reduced oxygen levels at higher altitudes involves slower responses which, despite the decreased supply of oxygen to the body, begin to restore its flow closer to that at sea level. Lack of oxygen releases an enzyme called erythropoietin (EPO), which stimulates production of new red blood cells in the bone marrow, increasing the number in circulation and boosting the blood's oxygen carrying capacity. Water then seeps from the tissues back into the blood. To compensate for alkalosis, the body excretes bicarbonate in the urine, slowly restoring the normal acidity of the blood. Meanwhile, the brain signals the thyroid gland to slow down metabolism, which conserves oxygen but decreases energy for heat or action. Most body functions are slowed in order to survive.

With acclimatization, the increase in breathing slows but does not return to normal. The loss of carbon dioxide is gradually reversed, but even after weeks at altitude, CO_2 blood levels remain somewhat lower than at sea level. The heart rate and cardiac output also return toward normal in a week or two. But with even slight exertion, breathing increases dramatically; at extreme altitudes, how much one can breathe is a limiting factor.

Given time, acclimatization is effective, but if we go too high too fast, the "struggle responses" may be inadequate and acclimatization too slow to provide us with enough oxygen, and we are likely to get "mountain sickness." Though we speak of several forms, these are really a continuum in which now some, now other signs and symptoms predominate. They are all a consequence of hypoxia.

About 20 percent of those who go rapidly from sea level to 9,000 feet or higher will suffer from headaches, nausea, sleep problems, weakness, or excessive shortness of breath, the unpleasant symptoms of Acute Mountain Sickness (AMS), which is usually mild and short-lived. The approach to Everest is gradual enough that few climbers suffer more than a mild headache from AMS.

AMS can develop into the more serious High-Altitude Pulmonary Edema (HAPE) when the small amount of fluid that appears in most lungs at altitude is not absorbed normally. Instead it accumulates, obstructing the flow of oxygen out of the alveoli and into the blood, in effect drowning the victim in his own fluids. Climbers suspect HAPE if the symptoms of AMS worsen and if an irritative cough appears, producing mucus that is often stained with blood.

A rarer part of the continuum is High-Altitude Cerebral Edema (HACE). The victim has trouble walking or using the hands; called ataxia, this condition is an early but ominous sign of a more serious brain disturbance caused by altitude.

Hallucinations appear and are often unrecognized as such by the victim. At this point the victim's life is in danger unless he or she descends to lower altitude immediately.

Supplementary oxygen and medications delivered at this stage are not always successful in reversing the condition.

Recently, we have hypothesized that the whole continuum of altitude-related illnesses is due to disturbances in the brain. The spinal cord is bathed in a clear liquid, the cerebrospinal fluid (CSF). Defending this fluid from the intrusion of unwanted materials from the blood are thin membranes, the blood-brain barrier (BBB). Studies suggest that even as low as 12,000 feet, hypoxia makes the BBB more permeable, allowing water and substances unfamiliar to the brain to leak into the CSF. This might mean that even common forms of altitude sickness are due to this leaky membrane, and could explain why dexamethesone (a steroid that suppresses this leak) helps combat mountain sickness.

Although symptoms of AMS are unpleasant, HAPE and HACE are dangerous. HACE can be subtle and more difficult to recognize because it weakens the higher functions of the brain—judgment, perception, memory, and will. Many high-altitude tragedies have been caused by the dulling of these faculties. Near the summit of Everest, every climber is in significant danger; at a time when clear thinking is critical, climbers are most impaired.

As climbers ascend, their bodies undergo physiological adaptation to reduced oxygen in the air. This process of acclimatization, however, may not keep up with the rate of ascent—leading to altitude sickness. Humans do not fully acclimatize to elevations above 19,000 feet or so—nearly two miles below Mount Everest's summit.

tired, make camp, spend time and energy making a meal, and after not getting any sleep you repeat the process in the morning."

ACCLIMATIZING DOES NOT MEAN ADJUSTING TO CLIMATE

Why all the moving up and down before the climb? The Sherpas and team are leapfrogging supplies from camp to camp, but they are also acclimatizing. Moving too high too fast can strain this process and lead to high-altitude illnesses.

Acclimatization is time-consuming. Most experienced Himalayan mountaineers recommend ascending no more than 2,000 feet a day, in order to give the body time to adjust to lower pressure and oxygen levels.

It's remarkable that humans can adjust to altitude to the extent they do. Without supplemental oxygen, a person living at sea level would collapse within a half hour and die soon afterward if suddenly taken to 20,000 feet. At the summit of Mount Everest—where the available oxygen is a third that at sea level—the same person would lose consciousness almost immediately and die within minutes.

The 3,000 vertical feet from Camp IV to the summit is supremely difficult. At extreme altitudes, physical performance decreases at an accelerating rate. Even though Everest is only

Not all the comforts of home—but most of them—surround team members in the Everest *team's Base Camp dining tent. Generators and solar panels power the fluorescent lights and a "communications corner" that includes phone, fax, and radios.*

ARE SHERPAS DIFFERENT?

BY CYNTHIA M. BEALL, PH.D.,

S. IDELL PYLE PROFESSOR OF ANTHROPOLOGY, CASE WESTERN RESERVE UNIVERSITY

Ever since Tenzing Norgay summited Everest in 1953, Sherpas have come to symbolize the capacity for extraordinary physical performance at high altitude. There are several tantalizing partial answers to the question of how they do it.

One factor is high-altitude residence. Most of the 3,000 to 4,000 Sherpas in Khumbu live at high altitudes, as their ancestors have for centuries. They start with a lifelong adaptation to high-altitude hypoxia. For example, a Sherpa born and raised at 13,000 feet has adapted to a partial pressure of oxygen 37 percent lower than that at sea level. Thus, the summit of Everest, where the partial pressure of oxygen is 67 percent lower than that at sea level, represents a smaller, less stressful change from everyday life for a Sherpa than for a native of low altitudes.

Their physiology exhibits several adaptations that allow Sherpas to excel relative to sea level natives. They breathe slightly larger volumes of air per minute; they have healthy, brisk hypoxic ventilatory responses, and they maintain slightly elevated concentrations of hemoglobin, the blood molecule that carries oxygen. Also, Sherpa muscles contain dense capillaries, which may enhance oxygen flow to working muscles, and Sherpa hearts use a more oxygen-efficient mix of metabolic fuel than do hearts of people from sea level.

Another factor in Sherpa performance is physical fitness, measured in terms of VO_2 max, the highest oxygen uptake an exercising individual can attain. Physically active Sherpas have a VO_2 max comparable to that of low-altitude athletes. The climbing Sherpas appear to be a select subgroup, with excellent fitness. Further, studies have shown that Sherpas maintain body weight and sleep soundly during a climb; thus they avoid two common and debilitating problems experienced by climbers coming from low altitude.

Some Sherpas may also have a genetic advantage if they possess the major gene, detected in another high-altitude population, that increases the percent of their hemoglobin carrying oxygen.

The combination of these factors has enabled some Sherpas to partially counter the reduced availability of oxygen in each breath, and to excel at high-altitude climbing, performing as well or better than the small elite of low-altitude mountaineers.

High-altitude performers, Sherpas have the advantage of generations of adaptation and acclimatization to low-oxygen heights.

777 feet higher than K2—representing a difference in barometric pressure of only 3 percent—there is a disproportionately much greater drop in the body's ability to obtain and utilize oxygen.

Everest is in a class by itself. Houston stresses that "the Everest summiter must climb the last thousands of feet relying on courage, determination, and drive, rather than on more acclimatization."

Indeed, after weeks above 20,000 feet, additional gains in terms of acclimatization are marginal. In Operation Everest, Houston found that humans deteriorate above this height even in the comfortable conditions of a decompression chamber. Dr. Peter Hackett, director of the Denali Medical Research Lab and an expert in high-altitude physiology, describes the slow degeneration that the team faces as "a slow death by starvation, dehydration, suffocation, and exposure."

Although they need to climb high for acclimatization, the team must limit their stay there. Much of the recent success on Everest has depended upon climbers fine-tuning this balance: acclimatizing just enough to summit before they are too debilitated to do so.

Individuals under similar conditions respond differently to altitude, but this is difficult to predict, Houston says. Those whose bodies naturally cry out for more oxygen as they ascend actually acclimatize better than those whose bodies don't react as much to the decreased oxygen.

FOLLOWING PAGES:
"Like tiny ants in a world made for giants," Edmund Hillary wrote of his team's experience on Everest. Here, climbers scramble over and around ice chunks the size of houses, seeking a viable route through the Khumbu Icefall's ever changing, ever dangerous maze.

"We know little about why this is so," Houston says. "but the major factors determining one's response to high altitude are: the speed of ascent, the altitude reached, one's health at the time, and genetic or other influences.

"The fact is, we don't know how best to acclimatize," Houston says with characteristic candor. "On the one hand, thousands of mountaineers have used the time-honored method of climbing slowly upward and descending to sleep—the siege tactics known as 'climb high, sleep low.'"

"Alpine style" is the other method, in which climbers live at Base Camp and scramble up nearby peaks, going a little higher each day, getting tough and acclimatized. When the weather and individual conditions match, they go for the summit in one shot. "Dozens of people have summited both Everest and K2 from Base Camp in this style, too," Houston adds, "but some have died. It may be a riskier technique."

Whether climbers suffer brain damage from the hypoxia of a prolonged stay at altitude is hotly debated. Some have, the physiologists note, but usually these victims also experienced severe altitude illness, hypothermia, exhaustion, or trauma.

The team was fully aware of Houston's straightforward advice on acclimatzing: take time to go high, don't overexert or overeat, drink extra water, and listen to your body.

And, as Jamling reminded them, listen to Miyolangsangma, the Maiden of the Wind.

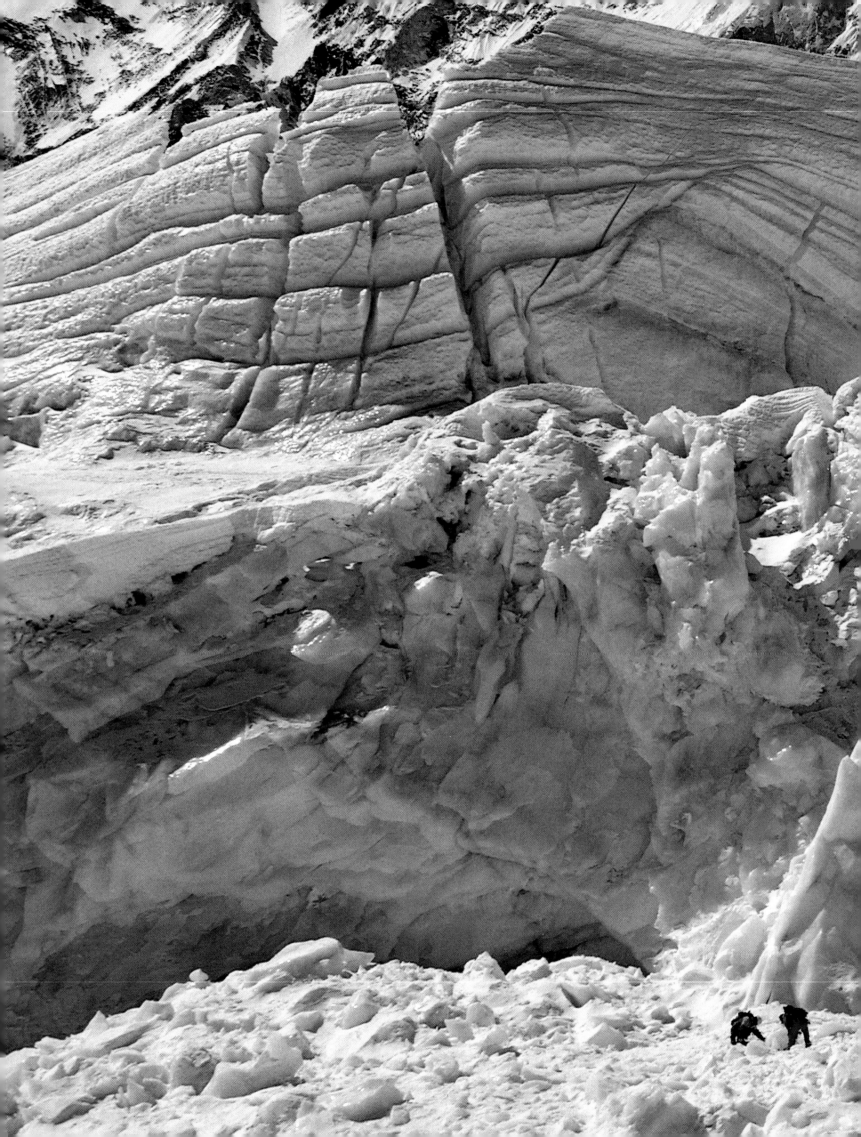

CHAPTER FOUR

THE MOUNTAIN THAT ROSE FROM THE SEA

"…it is the wind which adds to the sense of dread which possesses this place."

—SIR JOHN HUNT, REFERRING TO THE SOUTH COL

Climbers first viewed the Khumbu Icefall from the Nepal side in 1950, the year that Dr. Charles Houston joined the first group of foreigners to visit Khumbu. The group's journals, he said, consisted largely of superlatives expressing Khumbu's beauty and grandeur. But their glimpse of the Icefall itself was sobering. ¶ "For a long time we looked at the terrible icefall coming out of the Western Cwm, and decided that the approach would be very dangerous and difficult, perhaps impossible." ¶ This maze of crevasses, seracs, and ice blocks the size of apartment buildings has claimed more lives than any other part of the mountain. Avalanches routinely thunder down from the Lhotse wall and Everest's Southwest Face, and wash over the glacier; chunks of ice weighing hundreds of tons shift and tumble without warning; seemingly

Standard tools of the alpinist's trade, fixed ropes and steel-toothed crampons enable a climber to walk up ice walls—in this case, one of the endlessly varied, sun-burnished steeps of the Khumbu Icefall.

bottomless crevasses open and close, albeit more slowly. Looking up at the Khumbu Icefall from Base Camp, one feels about to be deluged by a tidal wave of gigantic ice cubes.

The 1963 American Everest expedition, sponsored in part by the National Geographic Society, experienced the first death of a foreign climber on the Nepal side of Everest. Dr. Gil Roberts was leading the second rope through the Icefall, 10 yards behind the climbers ahead, when a 30-foot freestanding wall of ice fell over and entombed his good friend Jake Breitenbach. Gil was partly buried, but freed himself and began digging out others.

With great emotion, Gil had to cut the rope that led to Jake, hopelessly buried beneath the serac. Then, as he helped the injured survivors descend through the Icefall, he heard a cry that sounded nearly human. It was a gorak, the bird believed to be the auspicious bearer of human souls, flying up from a crevice in the ice next to them.

Jake's body emerged from the Khumbu Glacier in 1970, and a fellow team member, Barry Bishop, took him for burial on the ridge beyond Tengboche.

Bishop had summited with the '63 expedition, and later served for many years on the National Geographic Society's Committee for Research and Exploration. He and his son Brent were the first American father and son to both summit Everest. In 1994 Bishop died in a car accident en route to receive an award from the American Himalayan Foundation honoring this

Barry Bishop and Lute Jerstad traverse the Lhotse Face in 1963, with the Geneva Spur immediately ahead. Beyond lies the South Col and Southeast Ridge, which they would scale after overnighting at their team's high camp, at 27,500 feet on the Southeast Ridge.

achievement. At Tengboche, some of Bishop's ashes are now buried alongside Jake Breitenbach.

A SCARY KIND OF PLACE

The team would need the last three weeks of April to establish and stock the upper camps. But first they would need to get through the Icefall.

Customarily, the route through the Icefall is found and maintained by a single team, which is paid for this service by the other teams. In early spring before most expeditions had arrived at Base Camp, an international party led by Scottish climber Mal Duff explored the Icefall for the safest and most direct route through it. They were followed by "rope teams" carrying ladders, 6,000 feet of rope, and plenty of hardware: 100 ice screws and 100 pickets (one-meter-long aluminum stakes), used to anchor the fixed ropes and the ladders over the crevasses. The danger from falling is greatly reduced by the use of fixed ropes, which are secured along exposed sections of the route for the duration of the season. Climbers have an ascending device, or carabiner, connected to their harness, and they clip this to the rope; they slide it up the rope as they climb, and the carabiner stops them if they fall.

More than 60 eight-foot aluminum ladder sections can be used in the Icefall. For a horizontal bridge, up to four sections are bolted together or lashed with plastic rope. Over the middle of the crevasse, many climbers lower to a crawl because of the unnerving bounce.

"A ladder like that is a little *wow*," Araceli recounted excitedly. "I look down, and then at my small rope—if I fall someone will catch me and take me out, I hope." Before her first Everest attempt last year, Araceli climbed some daunting vertical routes in Yosemite. By comparison, the Icefall was a snap, technically, but it gave her pause. "There's a place where sometimes you see your friends over there and get near them, on the same level, and then you go down again," she added. "It's a labyrinth of shifting turquoise ice. This is the challenging and fun part of the climb, but you must be there at not the wrong time."

Ed recalled a precarious stretch he encountered in '94. "To get up one ice headwall we used nine ladders tied end-to-end, with guy-lines leading off to the sides like trapeze wires. That was a crazy ride—it curved over backward near the top, and swayed horribly." The lower side of the crevasse dropped daily, and each morning Ed and his team found the topmost ladder another five feet below the upper lip. Each day a Sherpa would strap extra ladder sections to his back, crawl up the tottering construction, and lash them to the top end.

Once the Icefall ladders are fixed, they require daily maintenance to accommodate movement, especially later in the season. Screws melt out, the glacier moves, crevasses widen, and sometimes whole sections of the glacier collapse and take out the route.

"Climbing through the Icefall is like trying to cross a busy interstate at night dressed in black clothes," wrote Jim Litch, a climber, and a doctor at the Himalayan Rescue Association's Pheriche Clinic. "The difference is that instead of doing it

once, climbers make the trip over and over to the point that it becomes an acceptable risk."

Remarkably, no one has died in the Icefall since 1992. On April 8 when Ed first climbed through it, he remarked that it was in the best shape and location he had ever seen it, due partly to the heavy winter and spring snowfall. Nonetheless, climbers travel through it only in the early morning while the ice bridges and seracs are more solidly frozen.

As summit day approached, the Sherpas climbed empty from Camp II to Camp III on the Lhotse Face, picked up loads and carried them to the South Col, then returned to Camp II the same evening.

Like the climbers, the Sherpas enjoy the challenge and competition, and many are as strong as the best climbers. "People who climb Everest boast of their success, but few of them mention that 95 percent of the work—the grunt work—was done by Sherpas," Ed commented. "I've noticed that the climbers and clients who talk most about reaching the summit tend to be those who avoid the day-to-day work needed to get them there. They're often the ones who get in trouble, too."

Ed doesn't ask the Sherpas to do anything he wouldn't. He works to create a rapport, cross-checks plans with them, and lets them know that he's not simply ordering them around. "I know I'll see them again next year, and I want to maintain a good working relationship—and to continue having a good time."

David and Robert created opportunities to film whenever they could. They set up the bulky camera in the Icefall, though they were unnerved by the conflicting demands of both

moving quickly through the treacherous landscape and waiting for the proper light. They got the shots they needed, despite one mishap. "Quick, a cloud is coming, we need to use this light!" David said to Robert during one precarious camera setup.

"I tried to concentrate, but somehow turned the loading dial the wrong way," Robert recalled. "The camera ran up to speed, but the registration pins weren't positioned properly and they perforated new holes in the film. The camera jammed, making a real mess of the film. It was like celluloid confetti."

Only a few days earlier at Base Camp, the film crew had been shooting a time-lapse shot of

Making history, Barry Bishop of the National Geographic Society glories atop Everest in 1963, a member of the first American team to make the summit. Son Brent repeated the climb some 30 years later.

golden clouds flowing past a silhouetted peak. The camera made an unusual noise, then began to smoke. Brad immediately thought of its magnesium construction: The camera itself could ignite. Acting quickly, he removed the lens, and a ring of electrical smoke blew out at him from inside the body of the camera. He traced the smoke to a faulty, inexpensive chip, a component they had been assured was bombproof. Fortunately, Brad Ohlund had brought a replacement with him, and the camera was running again in several hours.

The following day, several pairs of the team's crampons were stolen from the base of the Icefall, where they routinely left them while

walking about on the rubble at Base Camp. David called a meeting of the team leaders to alert them, to request their return should they turn up, and to discuss the meaning of the theft in the context of the long tradition of mountaineering etiquette. At camp later, he told Steve Judson that he felt the incident was a bad omen.

RESCUE

The Icefall crevasses are so deep that when Sherpas look into their blue-black depths they often joke that they're "looking into America." Falling into one is referred to as "getting a visa to the U.S."

On April 8, a radio call announced that there had been an accident in the Western Cwm. An unroped Sherpa on Rob Hall's team had fallen through a snow bridge and into a hidden crevasse between Camp I and Camp II. Miraculously, he had landed on a peninsular ledge surrounded by darkness and depth.

The Sherpa was pulled out by his teammates and was stranded at Camp I for two days with a suspected broken femur. The complicated evacuation involved 35 people, including six *Everest* team Sherpas, and took most of a day. From Base Camp the Sherpa was airlifted to Kathmandu. He was one of five sent by Rob Hall to establish Camp II before Hall had arrived at Base Camp and before Hall's Base Camp puja had been

An aluminum ladder—made from the more than 60 ladder sections set and maintained by a commercial expedition at locations throughout the Khumbu Icefall—bridges a crevasse for Sumiyo Tsuzuki. A fixed rope would stop her fall should she slip.

THE HIGHEST FAULT IN THE WORLD

BY KIP V. HODGES, PROFESSOR OF GEOLOGY, MASSACHUSETTS INSTITUTE OF TECHNOLOGY

Two fault types played an important role in the evolution of the Himalaya: thrust faults and normal faults. Thrust faults mark the zones along which large sheets of continental crust are shoved over other sheets.

If you've ever watched a snowplow at work, you've seen sheets of crusted snow piled up in front of the snowplow's blade. In much the same way, sheets of rock are piled up to form the thickened crust of a mountain range.

Normal faults develop in a different way: They thin rather than thicken the crust. Lay a deck of cards on a table and spread them out with your hand. This causes each card to slip past adjacent cards the same way that normal faults move, resulting in a wider but thinner stack of cards. Normal faults are usually found in regions of crustal thinning, like the East African Rift, rather than in mountainous regions. Their existence in the Himalaya implies that the range did not grow higher and thicken continuously throughout its history, but experienced intermittent episodes of thinning and "collapse." These probably occurred when the crust of the range became too hot and weak to support its own weight.

One of the most spectacular of the Himalayan normal faults is the Chomolungma detachment. It occurs near the summit of Everest but having never been studied by geologists on the mountain itself, its exact position is unknown. Based on aerial photographs, the fault is nearly horizontal, slicing downward through several of the peaks north of Everest and into the Rongbuk valley. Near the valley floor geologists have studied the Chomolungma detachment, and have learned that it separates Ordovician limestones above (between 434 and 490 million years old) from much older metamorphic rocks below. One of the rock units just below the fault is a band of yellow marble deposited as a limestone in the Cambrian period, then metamorphosed to marble much later during the Miocene epoch.

From near the summit of Everest, climbers have collected rocks similar to the Ordovician limestones in the Rongbuk valley. This, with other evidence, led geologists to speculate that the yellow marble in the Rongbuk valley is the same rock as the famous Yellow Band, one of the most conspicuous landmarks for climbers on Everest. If so, that would put the Chomolungma detachment somewhere between the Yellow Band and the summit, perhaps at the top of the Yellow Band. To help test this hypothesis, the *Everest* filming team collected rock samples within and just above the Yellow Band. The Yellow Band samples are indeed like the Cambrian marble at Rongbuk, but the sample collected directly above the Yellow Band is schist (metamorphosed shale) rather than limestone. This suggests that the detachment occurs still higher on the mountain. Samples brought back from other expeditions require the fault to be below the Hillary Step, but its exact position remains a matter of speculation. In any case, the Chomolungma detachment is almost certainly the "highest" fault in the world.

Long an eyecatching landmark for climbers on the way to the summit, the crumbly, colorful rock of the Yellow Band was laid down hundreds of millions of years ago as sediments lining the ancient Tethys Sea.

conducted. Some Sherpas connected the accident to this breach of tradition.

About two weeks later, Audrey reported that lights could be seen bobbing through the Icefall. Nawang Dorje, a Sherpa on Scott Fischer's team, had become sick and was assisted down by fellow team members on April 24. His symptoms of rales, caused by fluid in the lungs, resembled those of severe pulmonary edema, but when his condition didn't improve, doctors suspected it was complicated by other factors. Nawang remained in critical condition at the Pheriche Clinic, at 14,000 feet, until he was evacuated to Kathmandu by helicopter when the weather finally cleared enough to fly on April 26.

Audrey also reported that a member of Mal Duff's party had experienced a suspected heart attack on April 20 while climbing, and that he had been evacuated to Kathmandu. Ten days later, a Danish climber on Mal Duff's team was injured while descending from Camp II and suffered broken ribs. David and Robert were nearby, and found that the man could walk slowly and with difficulty. Robert and Thilen Sherpa elected to help him through the Icefall, where he was met by team members.

The team feared the likelihood of more accidents below Camp I. Many of the Taiwanese staggering through the Icefall didn't appear to belong on Everest: Some didn't know how to tie their crampons on and had difficulty walking in them correctly. The South Africans, too, were having problems. During the trek in, their three most experienced climbers resigned over concerns about safety, and described the expedition leader Ian Woodall as extremely unprofessional. For the remainder of the season, the South African team did not share the spirit of cooperation that is nearly always found in the mountains.

Henry Todd, leading his own group, was concerned enough to give ice-climbing instructions to South African Deshun Deysel, who had never been on snow and ice before. It turned out that Ian Woodall had never confirmed her permit as an expedition member.

David detected a certain look shared by the inexperienced clients: wearing the right gear, but out of place and awkward in it. He made a point of taking time to speak with the ones he met, urging them to be extremely careful and reminding them of the more dangerous parts of the climb. Guides for the commercial groups, aware of many of their clients' inexperience, suggested establishing a rule about turn-around time. After some discussion it was agreed that on summit day, if clients had not summited by 2:00 p.m., they would have to return to the South Col. No guide would want to gather clients straggling back to Camp IV in the dark.

On April 30, the team returned to Base Camp. Camps II and III were established. Camp IV on the South Col had been stocked with oxygen, food, and other supplies. They were as acclimatized as they would get, and now needed as much as five days to rest and to regain some of the weight they lost higher on the mountain.

One evening, a raucous party was held at Rob Hall's dining tent. Some climbers privately questioned why someone would have a party before the summit attempt. Others could understand the need to let off some steam; the combined tension of weather, logistics, crowding, and the dangerous mission of climbing the mountain had been building. For Hall, the stakes may have been raised: Neither he nor any of his party had reached the summit the year before, and this year he had two additional clients.

At dinner the team seemed relaxed. Each had reached Camp III and had returned safely. "The interesting part for me is the *missing mountain*—what erosion took away," Roger commented one evening with playful irony. "Everyone here seems awestruck by the stuff that was left behind."

"THIS ENTIRE RANGE LAY BENEATH THE OCEAN"

Even the summit of Everest contains rocks that were formed of sediments from an ancient sea—or seas. The 450-million-year-old marine limestone deposits on Everest may have been to the tops of mountain ranges and back to the sea more than once, and 10 million years hence the eroded remnants of Everest may again be on the bottom of the Indian Ocean. "In as soon as several thousand years from now—a mere geological instant—it is likely that Everest will cease to be the world's tallest mountain," Roger explained in an impromptu lecture that followed dinner one evening. "At some point, its top will succumb to gravity and simply fall down the slope in a big landslide."

Because they are so dynamic, the Himalaya are one of the earth's great laboratories for the study of mountain building and erosion. In Khumbu, the geology is immediately apparent, particularly above the tree line where there is little vegetation to conceal the cliffs and rocks.

Standing near Base Camp, Roger seemed to say, *Now it all makes sense,* as he turned and surveyed the stony mountainscape. Jamling saw him bend over to pick up yet another wayside rock.

"Put that down," Jamling gibed. "We may be carrying some of your scientific devices to the South Col for you, but don't ask us to start carrying *rocks!*"

Roger Bilham has helped pioneer the use of GPS (Global Positioning System) receivers to measure the movement of the surface of the earth, and he is working with the government of Nepal and local scientists to install a network of GPS survey points throughout the Himalaya. Using signals from GPS satellites, recorded over a week's time, the relative positions of these points can be measured to an accuracy of 3 millimeters. One permanently fixed unit was installed when the expedition passed through Namche Bazar.

The observations of vertical and horizontal motion are then combined with information about microseismicity and geological structure. "By mapping all of this, we can make inferences of the style and rate of deformation at depth, which in turn provide insight into the potential of future earthquakes," Roger explained.

Robert and David accompanied Roger up Kala Pattar, or "Black Rock," a relatively small peak overlooking Base Camp. Kala Pattar rocks are covered with a dark layer of "desert varnish" caused by thousands of years of solar radiation; when broken open, the rock is actually a light-colored granite. There, Roger activated one of the solar-powered GPS receivers, to measure and calculate the distances to Lukla, Pheriche, Dingboche, and Tengboche, where GPS readings were taken during the expedition.

The climbers would take a sixth reading on

Everest's South Col, an unwieldy but valuable location. Khumbu's valleys are covered with loose glacial moraine and their sides are too steep for the receiver. The relatively flat South Col has excellent sky visibility and is located on bedrock that is free of snow. "As long as someone else will haul the device up there, I'm delighted," Roger said.

The eight GPS navigational satellites broadcast on the same frequency, but to a stationary receiver their frequencies appear to shift because they are moving relative to each other—like the changing sound of a high-

Home away from home: Snow-shrouded tents of Camp II—also called Advance Base Camp— cluster on a bench at 21,300 feet.
The site is low enough that climbers can remain for extended periods of time without suffering the more rapid deterioration that occurs higher up.

speed train whistle when it passes an observer. The GPS receiver is programmed with satellite locations and contains a very precise clock, which allow the unit to distinguish between satellites and thereby calculate its location.

The GPS measurements showed that Namche Bazar is converging on Rongbuk monastery, directly north of Everest, at the rate of nearly a half inch per year.

"Our measurements indicate that the Indian plate is sliding northward beneath Tibet along an inclined plane roughly ten miles below Namche," Roger explained. "Above that plane, rocks scraped off the Indian plate are caught in the squeeze between India and Tibet, and are thrust upward.

"Ten miles down, points south of Namche that are temporarily glued to the Indian plate are being driven toward Tibet. But like a battering ram, Tibet is trying to stop the mountains near Namche from moving northward. Things are quiet to the south, but a lot of action starts to Namche's north. For example, the severe compression beneath Tengboche may be causing it to rise faster than Everest. According to recent data, Khumbu overall is rising about $1/4$ inch per year."

GREAT EARTHQUAKES ARE INEVITABLE

The rate at which the Indian and Eurasion plates are moving together appears to be fairly uniform across the Himalayan arc. And most of this convergence is happening within a region about 50 miles wide, on the edge of the Tibetan Plateau.

"Prithvi Narayan Shah, founder of modern Nepal in 1768, described Nepal as 'a gourd being squeezed between two hard rocks,'" Roger said. "He was expressing a political sentiment when he wrote this, but the analogy holds up just as well in the geophysical sense."

The evidence for this plate convergence is the many thousands of almost imperceptible earthquakes that occur every year in the Himalaya. Occasional strong quakes (Richter magnitude 7) signify major adjustments beneath the Himalaya, but many seismologists think that only the infrequent great earthquakes (those exceeding Richter magnitude 8) actually permit India's northward motion.

Where there are no great earthquakes, the plates aren't slipping, but rather are squeezing the Himalaya, "like the winding of a terrific spring," Roger explained. "The resulting buildup of elastic energy must eventually be released in the form of a catastrophic quake. In mere moments, parts of the Himalaya south of us may leap five meters—and possibly more than ten meters—toward the plains of India."

In theory, by measuring this strain using techiniques such as GPS, geologists should be able to estimate very roughly the frequency of great earthquakes.

In a given area of the Himalaya, a great quake can be expected every 200 to 400 years. Four of them have occurred in the Himalaya in the past hundred years, releasing high levels of strain that have built up across perhaps half the range. The remaining areas of the Himalayan arc where none have occured in a long time are destined to be next.

One of the largest earthquake history "gaps" lies between Kathmandu and Dehra Dun, in the Indian Himalaya. Scientists have been searching local archival documents and other accounts there for written evidence of quakes. They have found that in 1255 a third of the population of the Kathmandu valley, including the king, was killed by an earthquake. The king's son was crowned, and during his three-year reign several more earthquakes occurred—probably aftershocks, indicating that the original quake was likely a great quake.

"Since then, the news and historical accounts from western Nepal, which are limited at best, are oddly silent on earthquakes," Roger said. "Either the record was destroyed, or buried in a

WHERE THE HIMALAYA COME FROM

BY PETER MOLNAR, SENIOR RESEARCH ASSOCIATE, DEPARTMENT OF EARTH, ATMOSPHERIC AND PLANETARY SCIENCES, MASSACHUSETTS INSTITUTE OF TECHNOLOGY

Two hundred fifty million years ago, India was but a small fragment of Gondwana, a vast proto-continent that included South America, Africa, Antarctica, Australia, and other bits of land. By 100 million years ago, this huge landmass had started breaking into the smaller pieces of modern continents. Among them, India had begun its tectonic journey toward Eurasia, accelerating to the geologically breakneck speed of about ten centimeters per year, roughly four times as fast as fingernails grow.

The motion was slowed but not stopped when India's northern edge reached Eurasia's southern edge, initiating what is called the "India-Eurasia continental collision." The image of a high-speed sports car slamming into a parked 50-ton semitrailer truck, however, is misleading, if only because for 50 million years India has continued to pile into Eurasia with no hint of further slowing. The Himalayan Range is but surface damage, a bent fender compared to the widespread folding, fracturing, bending, and wrenching apart done to neighboring regions.

The impact of these two landmasses has compressed and uplifted southern Eurasia to create the Tibetan Plateau and, farther north, has plowed up a wedge of elevated terrain including the highest mountains outside the Himalaya and Tibet (the Tian Shan), plus the deepest lake (Baikal) and the second deepest basin on land

(Turfan). Scattered among them are graveyards of human disaster. The ongoing collision has spawned the world's most devastating earthquakes, several of which have taken more than 100,000 lives.

Prior to the collision, the ocean floor north of India had already been sliding beneath the southern margin of Eurasia. The lithosphere, the outer layer of the earth comprising both the crust and the strong uppermost mantle, carried as passengers not only the ocean floor but also the Indian subcontinent. The collision began as the Indian subcontinent was bent downward into a "deep-sea trench" south of Eurasia and, following the ocean floor in front of it, started to slide beneath Eurasia's margin. It might have been a steady, less eventful process, except for two important factors.

First, crust is light, and therefore buoyant; like a liferaft in water, it will not readily sink into the Earth. Instead of following the thin oceanic crust beneath Eurasia, India's continental crust resisted submergence and pushed northward, squashing and thickening southern Eurasia's crust as it advanced. Much of that thickened Eurasian crust now underlies the Tibetan Plateau and buoys up its high, flat surface.

Second, crystals of mantle rock appear to be stronger than those of chemically different crustal rock. At temperatures characteristic of the lower crust and uppermost mantle,

mantle rock can resist deformation, while crustal rock either fractures or flows like honey (on geologic time scales). Thus, as India penetrated Eurasia, slices of its northern edge were scraped loose and thrust back atop layers to the rear, while India's mantle and lowermost crust moved forward essentially intact. As it pushed southern Eurasia deeper into the rest of Eurasia, India carried this growing pile of crustal slices northward—a most impressive pile, for these slices almost entirely comprise the Himalaya. The boundary between the old southern margin of Eurasia and India's northern coast—the contact zone of the collision—is marked by the Indus and Tsangpo Rivers which flow through Tibet just north of and parallel to the Himalaya. All along these rivers, one can find scrambled blocks of dark green rock that once formed the floor of the Indian Ocean.

The Himalaya, however, consists of more than this. Deeply cut slices of inland, Eurasian rock were also sheared by the overriding offscrapings, then bulldozed by Tibet behind them, and eventually buried to depths of 20 to 30 kilometers or more. This pervasive shearing reveals itself throughout the high Himalaya by a consistent northward dip of layers, resembling a tipped stack of cards. This northward-dipping stack, 10 kilometers thick, occupies a zone 30 kilometers wide south of the high peaks. Warmed by the blanketing effect of

the thick pile of offscrapings from India, the buried rock was metamorphosed, and some of it melted to form the unusually white granite characteristic of recently melted Himalayan rock. This intensely sheared, melted, and metamorphosed inland rock was later exhumed to be exposed, as it is now, in high peaks and deep valleys. A particularly striking body of this notable rock underlies Nuptse, directly southwest of Everest.

Peaks rise at a rate of a few millimeters per year in response to the steady underthrusting of material. Periodically, avalanches and rock slides remove the top bit of rock, which rivers carry away as debris or mud. Mount Everest might rise a little higher each year for a while, but eventually a landslide or rockfall will lower its elevation several meters, to give some previously buried rock the chance to become the top of the world. The average elevation of the Himalaya doesn't seem to change.

While admiring the mountains, one might consider the amount of rock that Michelangelo removed in order to carve David. The shavings swept from his floor played an unsung, but critical, role in his creation. Glaciers and rivers have similarly sculpted mountains—works of comparable beauty—from the raw material of the Tibetan Plateau. Without the deep valleys, swept clean by the relentless scouring of erosion, there would be no peaks.

subsequent quake, or there hasn't been an earthquake there in a very long while."

The high accelerations (shaking) of great earthquakes can throw rocks out of the ground. With a gleam of youthful amazement, Roger explained that even algae growth on boulders can be used as an historic earthquake indicator: Great earthquakes can roll large rocks, exposing different sides to the surface and allowing new algae to grow, and this can be used to date the event.

EARTHQUAKE STUDIES

Seismologists are concerned about the next great earthquake. Because of the shape of the Himalaya, the seismic waves emmanting from the quake nucleus will be directed mostly to the south, and become focused within the gradually narrowing bands of Himalayan sediment in southern Nepal and northern India. As a result, the dense populations of these areas could experience a more severe jolt than points closer to the nucleus of the quake itself.

During the Great Bihar-Nepal Earthquake of 1934, the low mountains of eastern Nepal sprang southward almost six meters onto the plains of India. Tens of thousands died. If that earthquake were to recur today, the death and destruction would likely be much greater, partly because of the increase in population and

Documenting the Icefall's drama, David Breashears trains his lens on Araceli (above and opposite). The specially designed camera performed remarkably, despite the extreme cold, high winds, snowfall— and the wear and tear that comes with daily travel on Sherpa backs. A single, 500-foot reel of film lasts only 90 seconds, mandating few, if any, retakes.

partly because of the larger number of buildings, which are taller and poorly constructed.

Clearly, more than 100 million citizens of Nepal and northern India would benefit from even a limited ability to forecast earthquakes. Roger Bilham stresses the difficulty of accomplishing this. "We can't pinpoint the date of future earthquakes, but we can say that they definitely will happen; we can estimate their magnitude and acceleration and we can outline the the areas where they are likely to be nucleated."

Geologist Kip Hodges stresses that reliable forecasting may never be possible. One variable is that displacements on the surface may be mechanically decoupled from the deeper crust. Also, small continuous movements at one position may indicate a seismic creep, which lessens the probability of a catastrophic earthquake. Creep is the process in which rocks slide relatively smoothly, retarding the buildup of strain that causes severe earthquakes.

But Roger is confident that the earthquake silence in western Nepal is not due to creep. If the Himalaya were sliding smoothly over India, they would squeeze the hills and ranges of southern Nepal. But those ranges have not risen—at least in the last 20 years—to the degree that this process would require. Any creep should only delay an inevitable earthquake.

At present, only numerical probabilities can be assigned for the likelihood of earthquakes occurring within fairly long time frames, such as 30 years. Whatever advances are made, forecasting will never replace precautionary measures.

"Unfortunately, public officials are reluctant to believe that the recurrence of nightmare earthquakes is inevitable," Roger says, "and damaging moderate earthquakes don't occur with sufficient frequency to remind people of the dangers of poor-quality construction. So, wherever I go in the Himalaya, I publicize earthquakes, especially among urban planners."

ANTICIPATION

Like David and Ed, other expedition leaders were concerned by the number of teams on the mountain and the potential for a jam. To avert the costly and potentially dangerous consequences of unanticipated delays, Scott Fischer

and Rob Hall conferred on the timing of their teams' summit attempts.

Before even reaching Base Camp, Hall had wanted to try for the summit on May 10. In previous years he had gone for the top on this day, and felt it to be lucky: Nearly all his clients reached the summit in 1994, though no clients summited in '95.

Initially, Fischer wanted to try on the 9th. But when he discussed with Hall the shortage of time and the manpower needed to fix ropes high on the mountain, they decided to combine their experienced staffs. They chose May 10 as summit day for both teams, then notified the other expedition leaders, suggesting they plan around their date. Later parties, they explained, would benefit by the fixed ropes Fischer's and Hall's teams would leave behind on the tricky Hillary Step and the summit ridge.

David, Ed, and Robert had climbed Everest without fixed ropes above the South Col, and knew that Jamling, Araceli, and Sumiyo would feel confident doing so. They chose May 9, a day ahead of the two large groups. For the purposes of the film and for safety, David wanted the team to be relatively alone on summit day.

FROM BASE CAMP TO THE WESTERN CWM

On May 5, the team lit incense at the lhap-so and departed Base Camp for Camp II. "A part of me is excited and curious about what will

All but invisible against Everest, climbers head for Advance Base Camp at the glacier's edge. Up and to the right, the saddle of the South Col caps the upper reaches of the Western Cwm.

happen," Araceli said, "and another part of me is afraid. But this fear is what makes me respect the mountain."

David maintained a positive but cautious attitude. "We're right on schedule and if the weather holds, if we remain healthy, and if the camera doesn't malfunction, we should be able to get on top and get these images."

The climbers emerged from the Icefall at 19,800 feet just as the sun hit, and walked onto the massive glacial snowfield known as the Western Cwm (pronounced koom, a Welsh word that means "valley" or "cirque"). They deployed their umbrellas to block the powerful sunlight. The snow blanketing this low-angle valley acts like a giant solar oven, concentrating the sun that reflects off the walls of Nuptse, off the western buttress of Everest, and from the broad glacier valley that extends below them.

"When people think of Everest, they think of cold and wind. But some of the more uncomfortable conditions here are caused by excessive heat," David explained. "The ambient temperature in the Western Cwm can be well below freezing, yet you can't remove enough clothes to stay cool. We pray for a gust of wind, and when it comes our work capacity immediately rises."

Losing fluid through sweating—in addition to the large amount exhaled by heavy breathing—increases the risk of dehydration, which causes a wide range of symptoms and problems. "I'm trying to drink two to three liters of water a day, on rest days alone," Ed said. "When climbing, we should drink at least twice that, but at high altitudes we're often not thirsty. There's also the disincentive of having to get up in the middle of the night."

Camp II is set in the majestic, sculpted cirque of the upper end of the Western Cwm. The Southwest Face rises over 7,000 vertical feet above camp and seems to disappear into the stratosphere. Though temperatures at Camp II were still dipping below zero at night, the days were warmer than when the team first arrived almost a month earlier. In the afternoons, water flowed and gurgled over the rocks and made rivulets in the snow.

On May 7 the team left Camp II, and were slowed only by their moderate pace and by the bergschrund at the base of the Lhotse Face. Robert recalled that two weeks earlier, Yasuko Namba, who was not entirely confident using an ice ax, had to be pulled up the 50 vertical feet of the bergschrund by a Sherpa. At the ledge on top, Sumiyo met Namba, and the two nodded and bowed while they spoke politely. "They were clearly wishing each other the best of luck," Robert said, "but I could sense something of an air of competition between them."

The team climbed the steep Lhotse Face to Camp III, where tent sites had been carved out of the face itself. High winds continued to blast the Southeast Ridge far above them, and a mood of nervous tension prevailed. Their faces were swollen and sunburned, and they spoke little about the mountain.

Ed stirred a pot balanced on a small stove. "Well, we'll have some beef stroganoff for dinner, although Spam and Dijon mustard would be a more complete meal. Spam is high in protein, it comes out in convenient chunks, and you get 170 calories from only two ounces!" Araceli rolled her eyes.

Sumiyo turned the videotape camera on Ed,

and he conveyed a message to Paula. The cassette would be taken to Base Camp by a Sherpa the next morning. "We're heading up to the South Col tomorrow," Ed said confidently. "Please don't worry about me."

"I'm worried, myself," Araceli interjected quietly off camera, "mostly about being cold."

Sumiyo finished the taping and turned to Araceli. "Which socks will you wear for the summit?" She was aware of Araceli's problem with slow circulation.

"Oh, I can't decide between the red ones and the pink ones," Araceli feigned. She had commented earlier that modern climbing gear may be lighter and warmer, but that the mountain

Clouds ominously begin to run up the Western Cwm, dubbed the "Valley of Silence" by Swiss climbers in 1952.

FOLLOWING PAGES:

Long line of climbers with the Hall and Fischer teams threads out from Camp III bound for the South Col.

hasn't changed. "Your equipment and your body must be ready, but summit day will be the most difficult because you are weak even when you start out. It will be the other climbers, the team, and the teamwork that will determine whether we reach the top."

Three days earlier, Sumiyo had cracked a rib from violent coughing. She had cracked another one three weeks before that, while establishing the high camps. Though painful, little can be done for this injury. "My cough hasn't stopped, but my ribs seem to be improving," she said optimistically. The rest of the team looked at her with sympathy and hope, knowing how much she wanted to try for the summit and to install Roger's GPS and

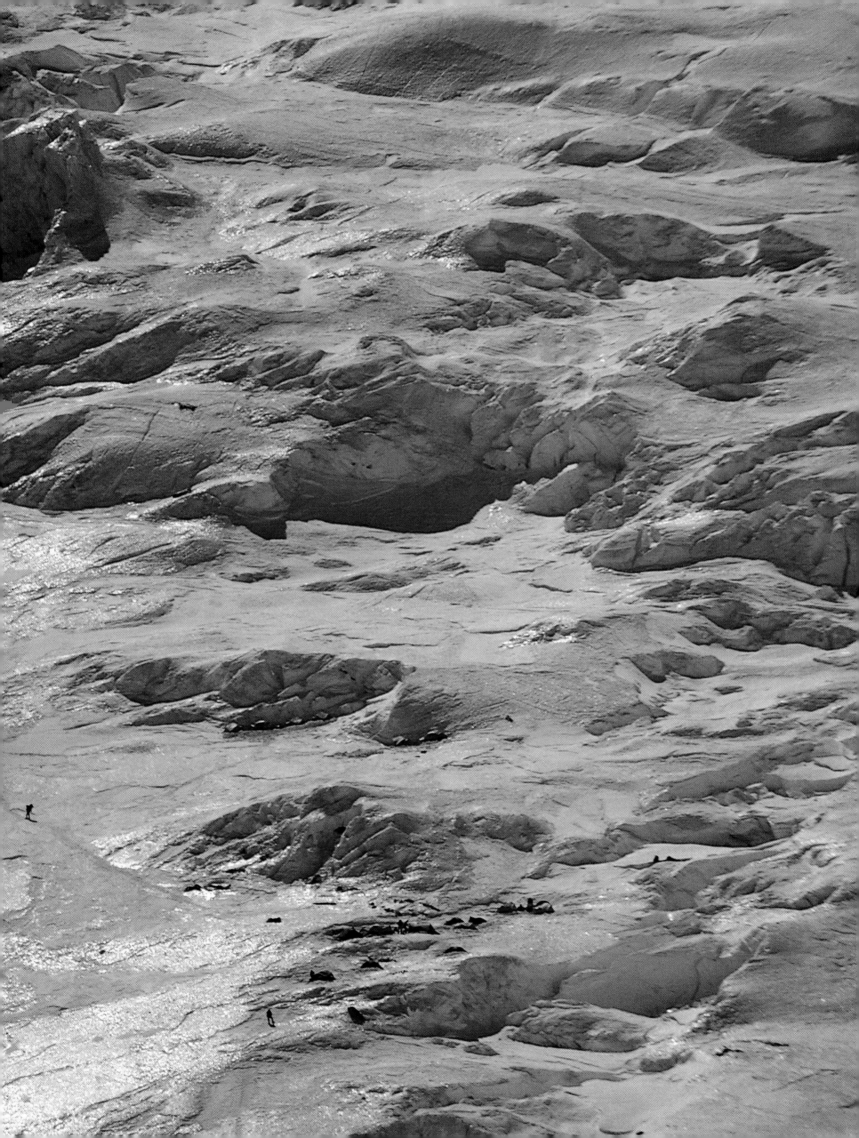

weather stations. "You know, I didn't tell my father that I was coming here to climb Everest," she added, sensing their concern.

"Now she tells us," Ed said.

"We're strong, and the rest days were good," Jamling said optimistically. "But I miss my wife, and my daughter will be ten months old tomorrow. I'll have to try to concentrate on the climb."

As they had done many nights, at Camp III David and Robert spent two hours cleaning and checking the IMAX camera and equipment. Robert worked long into the night loading film magazines with the bulky large-format film.

The team watched the upper part of the mountain. And listened. Fierce winds out of Tibet were still funneling over the South Col, threatening to drop into the Western Cwm. Robert called these gales, which sounded like freight trains, the "Lhasa-Kathmandu Express." Climbers caught in the middle of them describe the sound as varying between a deafening roar and an endless, nerve-rattling howl that flaps tents so violently that it's nearly impossible to talk. Sleep is difficult.

While establishing Camp III several days earlier, David saw a tent blow past him at 60 mph; on a previous expedition he had seen all the dining tents at Camp II get blasted away. "Your tent will be quiet for 15 seconds, then is suddenly barraged for minutes at a time," he grimaced.

The wind saps what small amount of warmth a hypoxic person is able to generate; Ed knew that summiting, particularly without bottled oxygen, would require a relatively calm day. Each year in May, there is a period of a few days when the high winds and unpredictable weather subside, presenting a window of clear and relatively calm weather. "That's the time to go for it," Ed attested, like a fisherman imparting a secret stratagem. "But if it turns bad, you have to be willing to retreat and wait."

For the climbers, weather is a gamble: It can start out fine and suddenly turn worse on climbs to the higher camps. And a bad day on Everest can be deadly. "There have been days in the past few weeks when we could have climbed Everest as mountaineers," David said, "but not as a film team. I have to be able to hold the camera steady and change the film bare-handed."

Robert had filmed in extreme conditions before, on a British expedition to the unclimbed Northeast Ridge. "The severe wind and weather kept us off the route for nearly two months, and during the summit push I had to turn back when my fingers froze while filming."

30 CLIMBERS, HEADING UP

That evening, the team looked down into the Western Cwm from Camp III and watched a stream of more than 30 people heading to Camp II. The group included two Taiwanese climbers and some Sherpas and guides, but most were clients of Scott Fischer and Rob Hall. Every one of them would be heading up to Camp III the following day.

Ed and David didn't need to speak. "We knew right then," Ed recalls, "that we did not want to summit on May 9." The descent from Everest's upper slopes is the single most dangerous part of the climb. The team would need to pass these climbers on their way down from the summit, and would have to unclip their carabiners from the fixed line and then reclip, many times, to get around people with unknown mountaineering skills. Would the clients step on the rope and cut it with their crampons? Dislodge some ice or rock?

"Their presence gave me a sense of being squeezed," David said.

Ed and David agreed that the team should descend to Camp II. They would wait for the others to "climb through." Besides, the moderate break in weather did not appear to be the traditional early to mid-May window of clearing, and waiting might also result in more favorable weather conditions.

On their way down on the 8th, the other groups ascended past them. David said he felt uneasy, as if he were bailing out on the mountain and his friends on other teams. Just above Camp II, Robert passed Rob Hall near the bergschrund at the bottom of the Lhotse Face. Despite the fact that the team was being extremely careful, Hall was concerned about falling ice or rocks that David and the team might dislodge onto his group. Robert felt that Hall might benefit by focusing his concern on the clients. Some of them were already on oxygen, though most climbers begin using it only above Camp III.

"The ankles of some were wobbling," Robert said. Not a positive sign. Shortly afterward, a rock that was likely set loose by an ascending climber whizzed by within a foot or two of Robert's head.

Araceli expressed the wisdom and experience of a true mountaineer. "Yes, when you pass people on the way down, you think maybe we didn't make the right decision— maybe tomorrow the weather will improve.

But turning back is never a bad decision. It gives you the chance to try again."

Another Catalan woman was attempting Everest from the North side, but Araceli said she felt no competition to be the first Spanish woman on Everest. "If you race when you are in the mountains, you lose the safe part of your mind," she said without hesitation.

A DEATH ON THE MOUNTAIN
Early in the morning of May 8, Chen Yu-Nan, a 36-year-old steelworker from Taipei, left his tent at Camp III on the steep Lhotse Face, possibly to relieve himself. He wasn't wearing crampons, and slipped 60 feet down the face and into a 15-foot crevasse. A Sherpa pulled him out.

Chen told "Makalu" Gau Ming-Ho, the leader of the Taiwanese team, that he would remain at Camp III, and assured him he would be all right. Believing his team member's condition to be fine, Gau departed for Camp IV alongside the Hall and Fischer teams.

Later in the day, the *Everest* team Sherpas returned to Camp III from the South Col. They found Chen in some distress, apparently suffering from internal injuries. They began to lead him down the fixed ropes. He was losing strength, but walking. At 3 p.m., the Sherpas radioed Camp II that his condition had deteriorated.

A half hour later they called again. Chen had collapsed about two-thirds of the way down the Lhotse Face, and they now believed him to be dead. They left his body attached to the rope and descended to Camp II.

Lying on the slope, Chen was visible from

EVEREST WITHOUT OXYGEN

The debate over the use of supplemental oxygen in extreme-altitude climbing began during the first attempts on Everest in the 1920s and has continued ever since. "When I think of mountaineering with four cylinders of oxygen on one's back and a mask over one's face—well, it loses its charm," George Mallory wrote. But his colleague, J.P. Farrar, an influential member of the British Alpine Club, countered: "Strictly speaking, I do not think that oxygen is any more of an artificial aid than food."

That mountaineers can climb "oxygenless" in the Death Zone highlights a coincidence of human and planetary evolution: The highest point on earth is also the highest point at which the human body can function without supplementary oxygen. Dr. David Shlim, Director of the CIWEC Travel Medicine Center, calls Everest's summit a "mystical point." If it were very much higher, the summit would likely be unreachable without supplemental oxygen.

At that altitude, simply surviving—not to mention climbing upward—is no easy task. "When you run out of bottled 'Os' at those altitudes, you might as well be underwater," observed climbing guide Pete Athans. "You retreat within yourself. You slow down."

Writer and Everest client Jon Krakauer felt that he was seeing the world "as if a movie were being projected in slow motion across the front of my goggles....You're dimly aware that this is cool and this is spectacular and that it's a long way down, but you're just sort of drugged."

Climbers are chronically hypoxic, and they report that the skin of other climbers has a bluish cast to it. "Your body turns into a machine to process whatever oxygen molecules it can find in the thin air," David Breashears noted. Reinhold Messner said that on his first oxygen-free ascent of Everest, with Peter Habeler in 1978, he felt he had become "nothing more than a single, narrow, gasping lung, floating over the mists and the summits."

Although our energy for metabolism comes from food, oxygen is the catalyst for the metabolic process, and its presence determines whether or not we can exert effort. "It's eerie," Ed Viesturs said. "If you turn someone's oxygen off, they stop moving. With oxygen you function better, think more clearly, stay warmer, and are able to climb steadily without pausing every step to suck in six or eight gasping breaths."

If oxygen is so precious, why do climbers climb without it?

"When I first attempt a Himalayan peak," Ed explained, "I climb without bottled oxygen, even if it keeps me from reaching the summit. My personal goal is to

Ed Viesturs looks forward to climbing all 14 of the world's 8,000-meter peaks— without "Os."

see how I can perform, to experience the mountain as it is without reducing it to my level. For me, how I reach the top is more important than whether I do."

Breathing bottled oxygen is not exactly like breathing at sea level. At the summit, supplementary oxygen effectively "lowers" one by only 2,000 to 4,000 feet, making a small but significant difference to oxygen-starved climbers.

"Once climbers are on oxygen," Ed continued, "they become stronger. But it's a bit of a crutch. Without it, I don't have a mechanical apparatus that can fail on me and thereby endanger me. The oxygen system is awkward. Sunglasses won't fit over the mask, so I have to wear goggles, which fog up. Also, I can't seem to suck enough air fast enough through the valves of the mask—I have to rip it off to take a full breath.

"Most importantly, I'm aware of the tricks that altitude and hypoxia can play on you. While climbing, I test myself, asking myself whether I'm aware of the conditions, of my actions, and of what is around me. Exhaustion and hypoxia can cause one to 'lose it' mentally, and I never allow myself to fall into this state."

"When I'm guiding, however, I always use oxygen. You're there for the clients, and oxygen does enable you to function better, both physically and mentally."

Camp II. David insisted that they check on his condition—there was a chance he might be unconscious. The weather had turned rough, but David, Robert, and Ed packed quickly and departed. When they reached the Taiwanese climber, they could barely see through the blowing snow. Chen was dead.

At least they could evacuate his body. The Sherpas would not use the fixed rope as long as a body was attached to it, although monks remind lay Sherpas that it can actually be propitious to see dead bodies, especially in dreams or when traveling. Nonetheless, many Sherpas are averse to seeing or handling them, which can bring defilement, and thereby bad fortune, especially if there is a "karmic link" to the deceased.

David, Robert, and Ed lowered Chen to the foot of the Lhotse Face and over the bergschrund. After packing him in a sleeping bag, they dragged the dead climber back to Camp II and left him nearby in the ice. He could be carried down by his teammates later.

When Ed radioed Makalu Gau on the South Col to tell him that his climbing partner was dead, Makalu did not seem to take in the news. He was so utterly focused on getting himself to the summit that he answered vaguely, "Oh...thank you very much."

"I was upset by Makalu Gau's decision to leave Chen and continue climbing, because his death affected me," David said. "I had to go up and bring down a dead Taiwanese climber. I had to close

his eyes and cover his face. I didn't like doing that, it was a senseless death."

It wasn't David's first experience with death on the mountain. In 1984, Dick Bass and David attempted Everest with the ill-fated Nepalese Clean-up Expedition. Team members Yogendra Thapa and Ang Dorje fell from the Southeast Ridge to the South Col. Sherpas stuffed their frozen bodies into a tent, but eventually the tent came loose and fell another 4,000 feet to the Western Cwm. Climbing together again in 1985, David found parts of their frozen bodies, which had shattered when they hit the bottom. He collected what he could and dropped the body parts into a crevasse.

"On the north side in 1986," David said sadly, "a Sherpa named Dawa Nuru was swept away by an avalanche. Ang Phurba and I found him, and several Sherpas helped us get his body to Base Camp. A lama was called, and he was cremated at the nunnery.

"When you see bodies, you don't have room for hysteria or emotions—it's a survival mechanism. But we do try to bring them down, to bring some closure for their relatives."

Jamling was troubled by the death of Chen, which occurred on the tenth anniversary of his father's death. At Camp II, he burned incense, prayed, and chanted. Those at Base Camp did the same at the lhap-so.

By the 9th of May, a Scandinavian, a Frenchman, two Spanish brothers, and Göran Kropp, the

FOLLOWING PAGES:

Volunteers on a mission of mercy, David Breashears, Ed Viesturs, and other stalwarts of the Everest *team brave a near-whiteout as they lower the body of fallen Taiwanese climber Chen Yu-nan down the Lhotse Face to Camp II. Meanwhile, the Taiwanese team leader, Makalu Gau, continued his ascent.*

155

lone Swede, had made unsuccessful summit attempts via the South Col route. Three of them had reached the South Summit, at 28,710 feet, or just below it, but were forced back by deep snow and high winds. Also, the Yugoslav team had tried for the summit on May 9, but abandoned their attempt above the South Summit. They arrived on the South Col at 7 p.m., exhausted. Audrey reported that one member ran out of energy a few feet from the tents and had to be hauled inside.

Like the team members and many others, Charles Houston is humbled by the mountain's size and austerity. "Man does not conquer a mountain any more than a mountain conquers man," he wrote in 1953. "For a few brief minutes, once in a million years, men have reached the summit of Everest and other high peaks. But how much more often have they been chased away, victims of bad luck, storm, or their own weakness. 'Surely the gods live here,' said Kipling's Kim, and he was speaking for many who have been awed by this magnificence."

Though this was Ed's eighth trip to Everest, he was approaching the mountain with similar humility, expressed in a practical way: "You don't assault Everest. You sneak up on it, and then get the hell outta there."

Clipped into fixed ropes for safety, climbers string out on the lower Lhotse Face near Camp III. Upon reaching the sprawling Yellow Band—a distinctive geological layer, visible here—they will head left toward the Geneva Spur.

CHAPTER FIVE

TRAGEDY STRIKES

*"The mighty summit…seemed to look down with cold indifference on me…
and howl derision in wind-gusts at my petition to yield up its secret—
this mystery of my friends."*

—NOEL ODELL ON THE DISAPPEARANCE OF MALLORY AND IRVINE

It was an hour past noon on May 10, yet Rob Hall's and Scott Fischer's teams were still high on Everest, striving toward the summit, knowing that descent in the dark was likely. Cold, exhausted, and hypoxic, the guides and climbers would also face the wind and rough weather that had been settling in on the mountain nearly every afternoon for the past month. ¶ By mid-afternoon, Base Camp had received radio calls from the summit. Audrey sat bundled at her laptop computer, and her cold fingers tapped out the day's news for the satellite fax: "Today…Rob Hall, with two other guides and three of his clients (Jon Krakauer of *Outside* magazine, a Japanese woman, Yasuko Namba, and Doug Hansen) along with three Sherpas, and nine members of Scott Fischer's group…reached the summit….We understand Makalu

Whiteout in the Western Cwm: Barely able to discern up from down amid rapidly worsening weather, Everest Film Expedition *members head back to Camp II with the body of a Taiwanese climber who perished on the Lhotse Face.*

Gau, leader of the tragic Taiwanese expedition, also made it up. However, all the ascents were around 2 in the afternoon, which is very late in the day. We are now anxiously awaiting news that all of the climbers make it safely back to Col tonight."

Incredibly, 23 people had summited. But at Camp II the response to this news was guarded. "They are making the top in middle of the afternoon? *Whooof!'* Araceli said, expelling air in surprise and worry. Ed scanned the radio, but overheard only one conversation: Scott Fischer was telling his climbing sirdar to inform Rob Hall that three of Hall's clients had abandoned their attempt short of the summit, below the Hillary Step.

Ed, David, and Robert turned and looked from Camp II down the Western Cwm, in the direction of Base Camp. In the distance, they saw a large cloud bank welling up. Around 4 p.m. it began to roll toward them. At the same time, another cloud bank enveloped the upper mountain, several thousand feet above.

At 4:30 p.m., Paula was making potato soup in the team's Base Camp kitchen. Geodetic engineer Dave Mencin came over from Rob Hall's camp and informed her that he had just overheard a chilling radio conversation between Hall and his guide Andy Harris. Hall, who was still above the South Summit, was yelling to Harris that their client Doug Hansen had collapsed and needed oxygen. Hall said he would be staying there to help Hansen.

This was clearly a life-and-death situation. Paula dropped everything, and when she emerged from the cook tent, she looked up and saw thick, dark clouds moving extremely fast into Base Camp from down valley.

"It was eerie; in two seasons at Base Camp, I'd never seen clouds like that," she said. "They were dark, purplish-black."

Locally generated clouds tended to roll up the valley each afternoon, but these were more ominous. When Audrey stepped outside, she was so astonished by their appearance that she called Changba and the other Sherpas to take a look. They took time to examine them, then pressed their lips together tightly. No one spoke.

It began to snow. Paula ran to Rob's camp, but heard no further communication from Rob or others on the upper mountain. She radioed Ed and David at Camp II.

They were stunned. Rob Hall had decided to stay with client Doug Hansen near the dangerously high and exposed South Summit, knowing that he couldn't get him down alone. Hall and Hansen were undoubtedly exhausted; their oxygen may have run out; and now they were struggling through a pitch dark whiteout, in a roaring, chilling wind.

"I tried to imagine what they were going through," Ed said. "A nightmare." High on the Southeast Ridge, the wind maintained a distant, menacing howl. David and Ed left the radio on.

CONFUSION ON THE COL

Some time after 8 p.m., Paula again radioed Ed and the team at Camp II. Information from Camp IV was sketchy, but Base Camp had heard that just a few of the 23 summiters and other climbers on the upper mountain

had returned to their tents on the South Col. Only those who descended quickly from the summit or who had turned around before the top had arrived at camp. No one knew where the others were.

David and Ed knew what it was like to return to the South Col from the summit. "You're exhausted, and when you find your tent, you crawl into it and assume the climbers behind you made it into camp," Ed explained. "You don't want to get up and check on anything, or even walk around—you just want to pass out."

A short distance from camp, a tragedy was unfolding. Not far above the Col, high winds

Closing in on the prize, Klev Schoening and Mike Groom labor up the final hundred feet or so to the top. Now all they have to do is get back down alive.

FOLLOWING PAGES:

As gale-force winds make their rope take flight, guides put in fixed ropes on the approach to the steep, rocky knob known as the Hillary Step.

and snow had begun to buffet the returning climbers. Because of the danger of downclimbing a steep section of rock-hard ice above the Col known as the "ice bulge" in the deteriorating conditions, one group of 11 climbers headed slightly east of the route they had ascended. By the time they reached the Col, darkness had fallen.

The Col is the size of several football fields and nearly as flat, resembling a vast, featureless plain. The raging storm limited visibility to a few feet, and when the climbers removed their dark-lensed ski goggles to see in the darkness, they had to squint into 60-mile-an-hour winds and driving snow. They were cold, hypoxic, tired,

hungry, and dehydrated—all of which reduced circulation to their hands and feet, and dulled their minds.

At Base Camp, members from every group gathered at Rob Hall's camp. Many had radios in their hands and were attempting to communicate with the climbers still on the mountain. The batteries on Scott Fischer's radios had died, and no one could reach climbers on his team.

"Our heads were spinning with unanswered questions," Paula said. "Our worst fears and best-case scenarios swam together in our heads. I slept half an hour that night. Mattresses covered the floor of Rob's communication tent, where 15 people were curled up like sleeping dogs, dozing fitfully."

The tent was mostly silent except for some sporadic crying and praying. The Sherpas stayed up playing cards. All were desperate for information, but those on the South Col responded infrequently and could give little information when they did. Helen Wilton, Rob Hall's base camp manager, made a list of names and checked them off as climbers were reported safe. Her list showed 17 people unaccounted for, a number that wouldn't be clarified until the next morning.

"Rob's situation was the only one we really knew about," Paula said, "and all night I'd been thinking that he would never stay behind with someone and not have somehow tried to *move*. He knew the danger of remaining stationary at that altitude." Every climber within radio contact had

All eyes on the mountain, climbers in Camp II react to the ominous news just radioed in: Taiwanese climber Chen Yu-Nan has suddenly collapsed while descending the Lhotse Face.

exhorted Hall to come down, assuring him that a rescue team could be sent for Doug the following day.

"It was shocking," Audrey recalled. "Here was a situation where a guide would have to leave behind a still-living client. I remember thinking it would finish him as a guide, and that he was in an impossible situation— damned if he did, doomed if he didn't. Even more shocking was the total lack of news about the 17 climbers who should have returned to their tents on the South Col."

After midnight, Mike Groom and Neal Beidleman, guides for Hall's and Fischer's teams, stumbled into Camp IV on the South Col with two of Fischer's clients and two Sherpas. In a state of near-total exhaustion, they described to Anatoli Boukreev, a Russian climber who was guiding for Scott Fischer, where the other climbers were huddled: Cold, exhausted, and unable to see, five more remained out near the Kangshung Face on the far eastern side of the South Col, about 400 yards away.

Groom, Beidleman, and the others barely made it to their tents. Boukreev, who had returned from climbing to the summit without oxygen, was exhausted, but he went out for the stranded climbers. Unable to find them, he returned for better instructions from Beidleman and Groom. On the next trip he found three clients, not far beyond the point he had reached earlier, and he guided them back to Camp IV. Two were left behind. Boukreev said that he

didn't see Beck Weathers. Yasuko Namba, if not dead, was unable to walk. Both were clients of Rob Hall.

THE NEXT MORNING AT BASE CAMP

At 4:45 a.m., Helen's radio went off like an alarm. "Is someone coming to get me?" the crackling voice asked. It was Rob Hall. Helen flew to the radio.

"Doug is gone," Hall said cryptically. His voice was partly muffled by his oxygen mask. It wasn't clear if Hall meant that he had been separated from client Doug Hansen, or that Hansen was dead. Hall, at least, hadn't moved from the South Summit.

Paula called Ed at Camp II and told him to get on Rob's radio. It would be best if a fellow climber, a friend, talked to him. Someone would *have* to get Rob moving.

Ed and other climbers gathered in Hall's

dining tent at Camp II. Several times Hall asked "'Where's Andy?—he was with me last night!'"

"We didn't know what he was talking about, because Hall's client Jon Krakauer had said that he had seen Andy Harris much lower down, on the South Col," Ed recalled. "Andy must have been up there with Rob and Doug for at least part of the night. Rob didn't give any further clarification about Doug, either."

Ed believes that Hansen could have fallen from the traverse between the Hillary Step and the South Summit, but that it is more likely he made it to the slot Hall used as a refuge near the South Summit, and they spent the night there together. Doug must have died during the night.

A DIFFICULT PHONE CALL

It wasn't until mid-morning on the 11th that people at Base Camp and Camp II were updated with news from Camp IV: Andy Harris was missing. Scott Fischer and Makalu

Gau were last seen below the Balcony, the ledge at the bottom of the Southeast Ridge. "Two bodies" had been identified outside Camp IV, those of Beck Weathers and Yasuko Namba, who were lying on the South Col.

At least two Sherpas and a doctor on Hall's team had hiked out and removed the snow from Yasuko Namba. One account said that her pupils were dilated but that she was still breathing, though barely. Another said that she was completely inert. They found Beck Weathers to be on the verge of death, and the doctor may have deferred to the Sherpas' judgment that the two could not be revived. They returned to the tents at Camp IV, and it was reported to Base Camp that Weathers and Namba were dead.

At Base Camp it was agreed that Beck Weathers's wife should be notified before she found out through the press. Helen Wilton phoned Hall's office in New Zealand. The office manager called Peach Weathers in Dallas and told her that her husband had been reported dead, and that his body had been identified. It was just after 7 a.m., Dallas time.

In addition to Hall and Hansen, Scott Fischer, Andy Harris, and Makalu Gau had to be somewhere on the upper mountain. But throughout that morning the Southeast Ridge and South Summit remained wrapped in clouds and furious winds, with a wind chill well below zero. The Sherpas at Camp IV were exhausted; getting them to head out from the South Col on a rescue effort took some serious

One of the most desolate campsites on earth, the South Col—location of Camp IV—boasts a grim mix of boulders, bullet-proof ice, and nearly constant hurricane-force winds—all at 26,000 feet.

coercing by Wongchu, who was at Base Camp.

Wongchu was also the sirdar for the Taiwanese team, and their Sherpas were on the South Col. Finally reaching them on the radio, he scolded them for staying in their tents. "Get up there, *now,* and look for those people," he barked, trying to sound intimidating. "Remind any Sherpas who remain in camp that I'm going to penalize them the moment they set foot in Base Camp." Ang Tsering, Hall's sirdar, similarly goaded another group of Sherpas into heading up.

Wongchu later admitted that while on Everest the year before he had dreamed of a goddess; she was smiling, and approached him and hugged him. In the spring of '96 the dream recurred, and the goddess again smiled and approached. This time, however, she turned angry, wrathful. During the climbing season, he mentioned the dream to no one.

DOWNWARD MOTIVATION
Paula was trying to be optimistic but finding it difficult. Even if the rescuers reached Hall and the others, how could they get them down? For the entire morning, she and Helen and Guy Cotter, a close friend of Rob Hall who was climbing on nearby Pumori when the rough weather rolled in, urged Rob to stay alert, to turn his oxygen up and get going. Caroline Mackenzie, Rob's expedition doctor, reminded him to knock the ice off his mask to get it to work properly. Oxygen would provide him the critical boost of energy he needed to start moving again.

In Hall's Camp II dining tent, Ed and others were also on the radio to Hall. "Several people were saying to Rob, 'Don't worry—we'll get to you!' as if trying to comfort him," Ed said. "But he was far too high on the mountain to be rescued. Paula knew this too, and she reminded me to be tough with Rob, to motivate him in any way possible. Rob *did* need to worry."

It may have been physically impossible for Rob to stand up and start moving. He would have been uncoordinated and weak after his night on the South Summit, and could easily fall off the mountain. Also, before he could drag himself downhill he would have to climb up and over the South Summit, a 25-foot rise. In his condition, this would be a monumental task.

Again and again, Ed pleaded into the radio. "Rob, you've got to get up and go—turn your oxygen on full, then crawl and pull your way up the rope and over the South Summit." He told Hall not to wait for the Sherpas, but to meet them halfway. He tried joking with him. "I told Rob, 'You have to come down at least for your wife and new child, who I hope will be better looking than you.' He laughed and appreciated that. For the most part he was coherent."

Veikka Gustafsson, a Finn on Mal Duff's team, cried during the radio calls, which went on for hours. He wanted to speak to Hall, but simply couldn't. He had first climbed Everest as a client with Hall in 1992, then soloed Dhaulagiri in '93 on another trip with him. Veikka, Rob, and Ed had climbed Makalu together.

By now, Ed and others were also in tears. Ed pulled himself together long enough to speak

with Rob, then broke down and turned off the radio so that Rob wouldn't hear him crying.

Helen Wilton called Dr. Jan Arnold, Hall's wife, in New Zealand, and held the phone to the radio. Camp II could hear the conversation.

"Rob sounded like a different person when he spoke with Jan," Ed said. "He became more lucid, and several times told Jan not to worry about him. He was positive and strong. That was when everyone listening in 'lost it.'"

"Okay, I'm going to try to get up to go now," Rob said finally. Everyone breathed a sigh of relief.

When Rob radioed again a few hours later, Ed was hopeful, upbeat even. "I said 'How's it going, Rob? Where are you now?' And he responded, 'You know, I haven't even moved—my hands are so badly frostbitten that I can't deal with the ropes.'"

David and Ed looked at each other, and the distress on their faces spoke for them. It was clear, now, that Hall would be unable to get up and move on his own.

COORDINATION FROM CAMP II

Others needed help. At Camp II the *Everest* team were preparing themselves for rescue efforts, but they first needed to communicate with Camp IV. Hall's tent at Camp II had a 25-watt base station radio, but the batteries of the one working unit on the South Col were dying.

The South Africans had radios at Base Camp and Camp IV, the only radios on the mountain capable of picking up all the frequencies used by other teams. They had climbed to the South Col on the 9th, intending to go for the summit on the 10th. But they had arrived on the Col exhausted, and had postponed their summit attempt.

Paula and others at Base Camp urged Philip Woodall, Base Camp manager for the South African team, to contact his brother, leader Ian Woodall, who was camped on the South Col. Perhaps the radio there could be made available for the emergency. Ian Woodall refused to help.

"They're on their own up there—exhausted, out of food, out of oxygen," Philip protested on behalf of his team. This didn't sound like a description of a team that was still considering a summit push, and it was clear Philip felt uncomfortable with his brother's decision. "You can't ask them to give up their radio, their one link with the outside world."

"It was crazy," Audrey said, "—the way the South Africans kept viewing themselves as separate from everybody else, as if on some other planet. If they had made an effort to join the community on the mountain, they would have realized that the community could benefit *them* as well."

David told Jon Krakauer to rip open the team's tent on the South Col and grab the batteries. The Sherpas had insisted that the tent be locked, following the theft of the crampons from the foot of the Khumbu Icefall.

With the radios on the Col now working, David wanted to assess the situation there before committing climbers and resources. If they were to rush up the mountain, they could get strung out and create an even more problematic situation. They needed to know exactly what kind of help was needed: clothing, food, oxygen, or rescue.

"I think everyone's right at their limit, now," David reported to Guy Cotter after speaking with Camp IV. "People are exhausted and hypoxic, and they're having trouble getting up and moving. There's a lot of will up there, but there's a lot of fear, too. We're trying to organize some fresh people to head up with supplies, but on our way up to Rob, for instance, how do we pass by Scott Fischer and Makalu Gau and others who may be in equally dire condition? This may be a triage situation."

CAMP IV IN SHAMBLES

Earlier that morning, May 11, Todd Burleson and Pete Athans, who had spent the night at Camp III, bravely set out for the South Col. The Lhotse Face was being hammered by high winds, and even those who stayed in their tents at Camp III feared being blown off the mountain.

As far as Todd and Pete knew at the time, 17 people were still unaccounted for, but near the Geneva Spur they passed a haggard and frostbitten group of climbers coming down. After receiving pieces of news about others on the South Col, Pete and Todd continued on.

Pete was appalled by the condition of Camp IV. "Our fears were confirmed," he said. "We saw gear strewn to the elements and the torn walls of tents flapping uselessly." They heard voices from one fairly intact group of tents. Inside, Jon Krakauer and another of Rob Hall's clients relayed what they knew. They said that the bodies of Beck Weathers and Yasuko Namba were lying at the edge of the Kangshung Face, several hundred yards away.

Pete and Todd went to work. David had

told them to take whatever they needed from the team's tents, which were filled with oxygen and supplies for their own summit attempt.

"David just gave freely of whatever was needed," Pete said, "with no idea whether he'd get anything back. Oxygen, for instance, is expensive to replace; without it, his team could lose its chance for the summit."

Everyone on the South Col was inside their tents, and most were too tired to get out of their sleeping bags—even to find nearby oxygen bottles. No one had been strong enough to assist anyone else, and there was no communication even between tents, though many were no more than five meters apart. The Sherpas—those who hadn't returned to the mountain to search for missing climbers—were exhausted, and some were disoriented.

"It was a scene from a horror movie," Pete recalled. "We had to get them down, but some of Rob's members didn't want to leave the Col without him. Only by busying myself with helping others was I able to suspend my incredulity and make the scene appear less surreal."

Pete and Todd encouraged those who were ambulatory and not overly tired to get off the South Col. "We began by delivering bottles to various tents, like pizzas," Pete said. They placed oxygen masks on several climbers, and gave them liquids and nourishment.

The climbers on the South Col were sinking into a dangerous condition of apathy that Todd and Pete found reminiscent of K2 in

Too high on Everest too late in the day, members of Scott Fischer's team descend from the Hillary Step around 4 p.m., amid a building storm. This photograph comes from Scott's final roll of film.

1986, when seven climbers trapped by a storm delayed too long on the Shoulder, which at just below 26,250 feet is about the same height as the South Col. When the weather eased, only three of them were able to make a break for Base Camp, and one of these died on the way. Five perished.

The winds were blowing a consistent 70 to 80 mph, with higher gusts, and a half-mile-long plume trailed from the Balcony. The Sherpas high on the mountain had yet to return to the Col, but Todd and Pete could see that it would be virtually impossible to rescue Hall.

As far as Todd and Pete could determine, the South African team remained in their tents the entire time they were on the Col.

THE SHERPAS TURNED AROUND

At about 4 p.m., Lhakpa Tsering and Rob Hall's climbing sirdar, Ang Dorje, descended to the South Col. They had climbed to within 800 feet of the South Summit—still two hours away in those conditions. They left some oxygen and a ski pole at the highest point they reached.

On the South Col, Ang Dorje sobbed as he recounted their efforts to reach Hall. Standing almost stock still while leaning into a wall of constant wind, tears streamed down his contorted face. Simply watching him was emotional for Pete, who was also despondent.

Todd radioed that one group of Sherpas had reached the Southeast Ridge, but had turned around in the battering winds.

David, Robert and Ed were climbing the

Lhotse Face to Camp III when the news about the Sherpas was relayed to David, who was a hundred feet behind Ed. Ed stopped and cried.

"At that moment all of us, I think Rob included, knew it was over for him. David told me that I'd better say good-bye. But what do you say—'hang in there'? The last words I had spoken to Rob were that I'd see him on the South Col. Perhaps it was best to leave him with that thought. I simply couldn't have said good-bye, even if the radio were in my hand."

"Our sense of hopelessness was profound," David said. "Rob was only 4,000 vertical feet from us, but we might as well have been trying to rescue the Apollo 13 astronauts. There's no force in nature that can get you up the Lhotse Face and Southeast Ridge fast enough, and there's no helicopter rescue at those elevations. Very few people have survived a night out above the South Col, and the common feature of the nights when they did was that the wind dropped."

After hearing of the Sherpas' abandoned rescue attempt, Hall kept his finger on the radio. David, Ed, and Base Camp could hear him crying. Guy Cotter told him that someone would try again tomorrow.

"Rob said that he couldn't last another night," Ed said. "He knew. But then he quickly turned it around and said, 'OK, I'll hang in there and be all right.'"

Intimations of mortality: Huddled in Rob Hall's dining tent at Camp II the morning after several climbers died and others remain missing, Ed and David radio Rob—still high on the mountain— entreating him to start moving down.

REFRESHMENT AT CAMP III

David, Araceli, Robert, and Ed continued climbing to Camp III. When they arrived, they worked with other guides there to set up a way station, and began heating water for cocoa and soup. Before long, the stream of exhausted climbers that Todd and Pete had passed filed down from Camp IV, some of them on oxygen. The team removed the climbers' crampons, got them into tents, and began warming and rehydrating them.

"I was looking at the face of despair," David said after speaking with several of them, "There was an uneasy mix of elation that they were alive and a sense that something terrible had happened." Many had minor frostbite on their faces. Ed and Araceli got them up and headed toward Camp II before they could cool and stiffen up; if one rests when exhausted, it is difficult to get moving again.

"During this time we were all focusing on Rob and the survivors climbing down," Ed continued. "But Scott Fischer and Andy Harris were out there too, and what were they going through?"

Shortly after the first group of Sherpas returned to the South Col, Ngawang Sya Kya and two other Sherpas also descended. They were assisting Makalu Gau, who was barely ambulatory and appeared to be badly frostbitten.

They had found Gau and Fischer not far from each other, little more than 1,000 feet above camp. Ngawang Sya Kya said that Fischer was barely breathing and that his teeth

were clenched and his eyes fixed, unblinking. He hadn't responded when they gave him oxygen and hot fluids.

The afternoon of May 11 was fading. Anyone alive now would likely be dead by morning, after a second night out above the South Col. Anatoli Boukreev began to gather oxygen bottles, aware that he was the only one with the strength to make another try at rescuing Fischer. Pete, Todd, and most of those on the radio from Base Camp were discouraging him, assuring Anatoli that no one would criticize him if he didn't go. Accidents and losses can be compounded when climbers are tired, anxious, and hypoxic. "When rescuing in extreme conditions," Todd said categorically, "you must become very objective—and very calm."

"And at the top of your list is self-preservation," Pete added. "If you don't make it down, neither will the victim."

But at Base Camp, Paula was trying to console Ingrid Hunt, Fischer's team doctor, who was sobbing hysterically, pleading for Boukreev to try to reach Fischer. Haltingly, Hunt described for Boukreev where a syringe of dexamethasone, a steroid that is beneficial for cerebral edema and believed to temporarily increase strength, had been stitched into Fischer's jacket. He should jab Scott in the leg with it.

Moments later, at about 4:30, Todd was standing outside his tent, speaking with Boukreev. Fifty yards away, to his total astonishment, he saw an apparition: A climber was staggering towards camp, straight into a 60-mile-per-hour wind.

"As I went toward him, I could see that his pile jacket was open down to his stomach, his eyes were swollen shut, and his arm was locked upright, parallel to his shoulder like a mummy in a low-budget horror flick," Todd said. "His face was so badly frostbitten that he was unrecognizable. Then I realized it had to be Beck."

Beck Weathers, the climber described as dead or near-dead by both a Sherpa and a doctor, had stood up and walked into camp. Todd and Pete looked at each other as if they had seen a ghost, then quickly put Weathers in Scott Fischer's tent, resigned to the fact that Fischer wouldn't return. Weathers's right arm was frozen solid and felt like a piece of ivory, they said, and he looked near death. They were

concerned about his chances of a heart attack: When a deeply hypothermic person is rewarmed, they are at risk of a fatal cardiac arrhythmia when the cold acidotic blood returns from the extremities to the heart.

Todd and Pete got Weathers into two sleeping bags and turned his oxygen to full flow—four liters per minute—then administered fluids and brought him hot water bottles. "Beck was exhausted to a degree that I'll probably never know," Pete said.

Todd transmitted the news about Weathers's reawakening to Base Camp. "I can believe anything, now," he said, "and if this guy lives, I bet he'll believe anything, too."

In Dallas, Peach Weathers received another phone call, again from New Zealand, four hours after the previous one. "I didn't register the part about him being in serious condition," she said later. "I just heard that he was alive, and I knew at that point I would see him again." Her brother had already boarded a plane and was flying from Atlanta to help with arrangements for Beck's memorial service. A wake had been turned into a celebration.

At Camp IV, Anatoli Boukreev was galvanized. "Now I'm definitely going up—Scott's alive, too," he announced.

"At that point, I could have believed it," Pete said. "I admit, when I first saw Beck I figured it must have been Scott."

Pete and Todd decided to stay with Weathers, while Boukreev packed some gear and headed up, at 5 p.m. He returned to Camp IV well after dark, and confirmed that Scott was dead.

The news was met with shock at Base

Camp. Ingrid Hunt threw her radio in despair, and it crashed on the glacial scree.

A FINAL RADIO CONVERSATION

Around 6:00 that evening, May 11, as Rob Hall was beginning his second night on the South Summit, he again talked to his wife. "Sleep well, my sweetheart," he said to her before signing off for the last time. Audrey said that from the tone of Hall's voice—and against all odds—people almost felt they would be speaking with him again in the morning.

"Our radios were still on, and we continued to talk to him," Ed said, "though we knew that to survive another night was next to impossible. Here was an intelligent, thoughtful person, realizing he was going to die. Earlier, he said he was shivering uncontrollably, but toward the end his senses were dulled and he was apathetic, which is what happens in extreme cold. He was not in pain. When he turned his radio off, he lay down and fell asleep. In those conditions, death is a relief."

A SECOND NIGHT ON THE COL

Only a few at Camp IV knew that Beck was alive. They presumed he would be dead by morning, but in any event would be unable to walk—which was as good as dead. Carrying him down from the South Col would not be possible. That evening, there was some confusion over who would be looking after him. Weathers remembers being alive on the South Col, but alone and in great distress. "During the night I had some water bottles, but because my hands were frozen I couldn't get to them, and I couldn't get anyone to help me. Then I noticed that my arm was swelling, but it was constricted by my cheap plastic watchband, causing loss of circulation in my right hand and lower arm."

At one point, a Sherpa poked his head in the tent, but his English was poor and Weathers couldn't communicate his desire for water or to have his watch removed. In the pitch dark, he unsuccessfully tried to gnaw the watchband off. In the meantime, gale-force winds were blowing the tent over on him, similar to what other climbers on the Col were experiencing, so he rolled on his side in order to create a space to breathe. Some time before sunrise, the tent door blew inward, and the tent quickly began to fill with snow. When it finally became light out, he began yelling for anyone who might happen to hear.

"Jon Krakauer stuck his head in the door, and he was surprised as hell," Weathers said. "I asked him if he might be able to get Pete Athans." That's when Pete and Todd prepared Beck to descend. It was the morning of May 12. Fortunately, the wind had subsided, somewhat.

HAVING ARISEN, HE DESCENDS

Beck drank a liter of water and sipped at some soup. Although he was having trouble seeing, he was able to walk—haltingly, and with assistance. Pete and Todd had assumed Beck's feet were frozen, and hadn't removed his boots because they wouldn't have been able to get them back on his feet once they had thawed and swelled. Unlike Makalu Gau, whose feet were badly frozen, Beck was wearing a new model of mountaineering boot insulated with high-tech materials.

A NIGHT OUT IN THE DEATH ZONE

BY SEABORN "BECK" WEATHERS

As I approached the Southeast Ridge shortly before sunrise, I was feeling strong, but my eyes simply weren't focusing. Fortunately, I didn't really need to see the route, because deep steps had been kicked ahead of me. The traverse at the bottom of the Southeast Ridge required more vision, however, and I had great difficulty feeling my way along it. When we reached the Balcony, I had to tell Rob Hall that I wouldn't be able to continue climbing—for the moment. In the brightness of the sun perhaps my pupils would constrict and I could follow later, I told him optimistically.

"Only if you're able to leave here within the next 30 minutes," Rob told me.

"Well, if I can't, then I'll just head back down the mountain."

But Rob didn't like the idea of not knowing whether I had made it down safely or not, so he made me promise to stay put until he returned.

I was still waiting there for Rob when the evening light started to fade. My vision again deteriorated when my pupils dilated. I now regretted my promise to Hall, especially because some hours earlier, around 1 p.m., others on our team had abandoned their summit attempt and offered to help me down.

Jon Krakauer, a teammate, was the first climber to return from the summit. He didn't mention having seen a storm coming, though in one of his accounts he reported that when high on the mountain he noticed that to the south a blanket of clouds had quickly replaced clear skies. I told Jon that I really couldn't see very well and that I needed to descend, and might need him to downclimb close enough to be my eyes.

Jon was willing to descend with me, but he reminded me that he was not a guide and that Mike Groom was coming 20 minutes behind him. Mike had a radio, and could let Hall know that I was heading down with him.

When Mike descended, he was assisting Yasuko Namba, who was badly exhausted. Neal Beidleman also came, with clients from Scott Fischer's group. Mike turned Yasuko over to Neal, then short-roped me down the Triangular Face.

From the face we climbed onto the South Col, and were there for only only a few minutes when the storm came up—very quickly. I was cold but not particularly tired, and held onto Groom's coat sleeve. Visibility went to zip, and in the blowing snow and gathering darkness the other climbers became nothing more than fuzzy, disembodied headlamps. Totally lost, we were a pod of people following, like kids playing soccer, whoever was the current leader. We came to a standstill within feet of the sheer drop-off of the Kangshung Face, on the eastern edge of the South Col, and formed a huddle.

All of our oxygen had run out, and we rubbed and pounded on each others' backs, trying to keep every muscle in our bodies moving in order to generate heat and stay awake. I removed my right mitten, while leaving on the expedition-weight polypropylene glove liner, in order to place my hand inside my parka to warm it. The skin on my arm instantly froze. In that instant the wind blew my mitten away, and suddenly I was unable to zip up my parka. The spare pair of gloves in my pack might as well have been on the face of the moon, and I couldn't have opened my pack anyway.

After a few hours, some stars shone through a hole in the clouds, and we had a halting discussion about how to proceed. Some of us were barely able to walk, and I couldn't see. Groom and Beidleman decided to strike out in search of the camp, to send people back for us. This seemed reasonable.

Gradually, the whole scene became more remote. I had a sensation of floating, and didn't feel cold anymore. That must have been when I drifted off. I was not conscious when Anatoli Boukreev returned for the others.

Some time the next afternoon, I found myself alone on the ice. I was not terribly uncomfortable, and was convinced I was dreaming: The hardest part was coming to grips with the fact that my situation was real, and serious. I rolled over and looked at my right hand, which appeared like an unnatural, plastic, twisted gray thing attached to the end of my arm—not at all the hand that I knew. I banged it on the ice, and it made a hollow sound, a sickening thunk.

This focused my attention. I could see my family there in front of my eyes, and managed to sit up, realizing that if I didn't get moving, I was going to lie there for eternity. None of our group was there; either they had left or I had become separated from them. It was clear that help wouldn't show up now.

I dumped my pack and ice ax, figuring this was a one-shot deal: I would either find camp or lose my last remnant of energy and sit down to wait for the end. For about an hour and a half I wandered in different directions, unable to orient myself, hoping I'd recognize something.

Then I remembered that during the night someone had said that the wind blows over the South Col from the Western Cwm, from the west. Camp had to be upwind. So I turned into

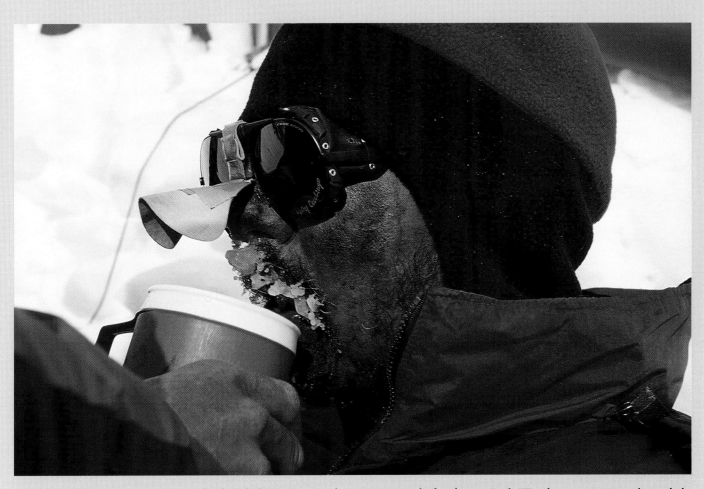

Left for dead not once but twice, his hands frozen into uselessness, a severely frostbitten Beck Weathers receives much needed fluids at Camp III after weathering a horrific night on the mountain, totally exposed, face down in the snow.

the wind, put my head down, and figured I'd either walk into camp or off the edge of the mountain.

I was propelled by a primitive desire to survive. My oxygen-starved brain wasn't working, but I was certain of one thing: that I would die, that very soon I would sit down in the snow and wait for exhaustion and the cold to overcome me. I began to hallucinate. The landscape was moving and the rocks changed shape and crawled around on me, but I accepted this and continued wandering. It was not at all frightening. I was in a very calm state, except for a feeling of sadness that I would be unable to say some of the things I wanted to to my family. I knew that I could accept death.

But I had a heck of a lot to live for, and wasn't going down easy. My family, standing there before me, became an enormous driving force. The changing, uneven surface of ice and rocks caused me to lose my balance and fall several times. I knew not to fall on my hands, so I rolled as I went down—which was exhausting in itself.

And then a miracle happened. A couple of soft, bluish rocks appeared in front of me, and their smoothness led me to think they might be tents. But right away I caught myself indulging this thought, knowing I would only be disappointed, which would affect my will to continue. I steered toward them anyway, preparing to walk right past them.

Suddenly, someone was standing there, and it was Todd Burleson. He took one look at me, got me by the arm, and led me to camp. Pete Athans and Todd were sure that I was going to die, too, but I'm glad they didn't tell me. When a middle-aged guy like me can survive that, it gives truth to the possibility that this kind of strength resides in each of us.

Pete radioed Ken Kamler, a doctor on his team at Camp III, that they were bringing Weathers down. Camp III consisted of a few narrow tent platforms shoveled into a 35-degree slope; there was no room to treat anyone. The day before, Kamler had requested that Base Camp send up medical supplies to Camp II, and after talking to Pete he downclimbed in order to set up a field hospital and waited for Beck to be brought down.

Some years earlier, Beck had undergone a radial keratotomy procedure to correct his eyesight. He was unaware that a potential side effect of this procedure was impaired vision in hypoxic conditions. The side effect had only recently been reported in medical journals, and several climbers who had undergone the procedure had experienced no vision problems while high on Everest. The effect, thought to be rare, is always temporary. Additionally, the general swelling of Beck's face from frostbite and exposure had further restricted his vision.

Todd and Pete gave Beck a shot of dexamethasone which they hoped might improve his strength, put a harness and crampons on him, and started on the traverse to the Geneva Spur. Climbing down the mountain, they had to describe each step for him.

As they left the Col, Pete and Todd looked back briefly, painfully aware that they were leaving their friends Scott and Rob on the mountain. "My emotions can't accept logic as consolation," Pete wrote later. "The image of two shivering, lonely figures, solitary and abandoned, frequents my restless dreams."

Robert and Ed started climbing above Camp III in order to help Pete and Todd with Beck's descent. They met them at the top of the Yellow Band, and the four of them lowered Beck and rappelled beside him. Once Beck was on the fixed lines of the Lhotse Face, he was able to downclimb. His arms were still frozen and he had little coordination, but he had regained partial vision in one eye, with some depth perception.

"Beck became very lucid," Todd said. "He knew he would lose his hands—but he joked about his great disability plan, for instance. When hypoxic, people are often aggressive, fighting you. But Beck was congenial, and he told us to simply 'do what you gotta do.'"

"Beck was quite funny," Robert said in amazement. "He wasn't that depressed. The pathologists in my country, too, are the ones who keep their good humor even in bad situations. In the Alps, I've carried many people down who looked better than he did, and they *died* on the way."

At Camp III, Weathers asked for black tea with sugar. Because he couldn't clutch anything with his frozen hands, he had to be fed.

On his feet but unable to see or to use his hands, Beck Weathers (third from right) eases down from Camp III with the help of climbers from three different teams.

FOLLOWING PAGES:

Rescue attempt at 19,700 feet: Flying near the altitude limitations of his helicopter, pilot Lt. Col. Madan K.C. homes in on a Kool-Aid "X" and Ed's makeshift wind sock just above the Icefall, hoping to evacuate Beck and the severely injured "Makalu" Gau.

"Guys, I'm gonna lose my hands, but I just might see my wife and kids again, if I can ever make it down."

"You can do it," David replied.

"If you think I can do it, I bet I can," Beck said.

At Camp III, David jumped in and helped with the descent of the Lhotse Face. At all times,

one of them walked in front, providing support and guiding his feet, while another walked behind, gripping Beck's harness. The anchors for the fixed ropes proved tricky; the climbers had to unclip Beck and then themselves from the rope, and then clip onto the rope again on the far sides of the anchors. Occasionally losing

"THEY WERE THE HANDS OF A DEAD MAN"

BY KENNETH KAMLER, M.D.

Quickly, I had to prepare to treat two critically ill climbers who were being led down to Camp II. They belonged in a modern hospital ICU but I would be treating them in the New Zealand mess tent, trying to work out complex medical problems at an altitude where tying your shoes can be confusing. With help from Sherpas, other climbers, and Dr. Henrik Hansen, we laid foam mats and sleeping bags on the tent floor, gathered dry clothes, hung IV bags through carabiners, laid out oxygen bottles and regulators, arranged bandages and medications, and boiled up large pots of water. While we waited, the Sherpas used some of the hot water to serve tea.

Makalu Gau arrived with all of his fingers and toes frozen. I used a scalpel to cut away a piece of sock stuck to his foot. He had the worst frostbite I had ever seen—until I saw Beck Weathers, who was brought in just as we were getting Makalu stabilized.

I had expected an incoherent, half-conscious phantom, but Beck walked in mostly under his own power. In an easy, conversational tone he said, "Hi, Ken, where should I sit?" He was alert and coordinated, showing no signs of hypothermia. We laid him down on a sleeping bag and replaced all his clothes, which were wet down to his underwear.

When I removed his oxygen mask, I was shocked. Edema had swollen his face to twice its normal size. His cheeks were black and his nose looked like a piece of charcoal. His right hand, a third of his forearm, and his entire left hand were deep purple and frozen solid. They radiated cold. There were no blisters, no pulses, no sensation, and no pain. They were the hands of a dead man, but bizarrely, he could move his fingers: The live muscles in his forearm were able to pull on the dead bones in his hand.

I started an IV and injected nifedipine, a drug that diverts blood flow to the extremities, but can cause a sudden loss of blood pressure. Having no pressure cuff, I had to monitor him by feeling the strength of the carotid artery pulsations in his neck. His hands were placed in tubs of water heated to 104° F,

but they were literally blocks of ice, and cooled the water rapidly. Maintaining the temperature required continually drawing off the cold water and adding hot water from a thermos. The Sherpas were eager assistants and quickly got the hang of it.

As I worked, Beck talked casually. If you had simply heard the conversation and not seen what was going on, you would have thought he had just dropped by for tea. Beck and Makalu were under control, but I stayed up all night with my two patients, changing their IV's, adjusting their oxygen, and watching them breathe. To prevent the IV bags from freezing, I wrapped chemical hand warmers around them. It was a long and miserable night.

At 7 a.m., we heard a discouraging radio message that the

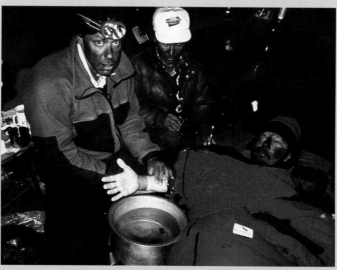

Climber and physician Ken Kamler treats Beck Weathers's frostbitten hands in Camp II.

rescue helicopter could not come above the Icefall because of the wind. This left me with a hard decision: either wait out the wind at an altitude where even minor cuts won't heal, or descend to Base Camp, exposing Beck and Makalu to further cold and trauma—and the rescuers to the dangers of evacuation through crevasse fields. I opted for the over-the-ice evacuation.

While rescue teams were being organized, I covered each of Beck's and Makalu's fingers with vaseline gauze, wrapped their hands in bulky boxing-glove type dressings, and then covered them in down booties to keep them warm. Once their oxygen masks were removed, I was able to treat their frostbitten faces by applying white Silvadene burn cream. I remarked to Beck that I was putting on makeup so he'd look good at Base Camp.

Beck could walk, but Makalu's feet were frozen and he had to be carried. I followed behind the two teams, down the Western Cwm, and was so engrossed in how I would manage them at Base Camp that I didn't notice that the wind had died down. I was startled back to the present by the sound of a helicopter overhead.

Kenneth Kamler has been on four Everest expeditions. He is a micro-surgeon from New York specializing in hand surgery.

his balance, Beck would reach for the rope, though his hands were unable to grab it.

"I used to tell people that I could do my job with my hands tied behind my back—now I'll have a chance to see if I really can," Beck quipped. He compared their tight group to a conga line, and the climbers joined him in a chorus of "Chain of Fools."

At the bottom of the Lhotse Face, Beck had to be lowered down the 40-foot vertical wall of the bergschrund. The climbers didn't need to discuss their effort. They had worked together long enough to know exactly what needed to be done. At Camp II, Dr. Ken Kamler took over, and was helped by Sumiyo, who had also cared for Makalu Gau when he descended.

EVACUATION

At Base Camp, Guy Cotter figured that getting a helicopter above the Icefall would almost be possible. But the atmosphere at 20,000 feet provides half the lift for takeoff that is available at sea level, and talking a helicopter pilot into landing near Camp I would be no easy task.

Cotter had phoned people in Kathmandu who were able to reach the U.S. Embassy. Simultaneously, a group of women in Dallas, mobilized by Peach Weathers, phoned their senators and the State Department. But by the time queries from the U.S. arrived in Nepal, the American consular officer in Kathmandu, Dave Schensted, had already alerted the Royal Nepalese Army of the need for a rescue.

"I had trekked to Base Camp two weeks earlier," Schensted said, "so I knew the lay of the land up there, which was helpful in discussing the logistics of getting a helicopter to Camp I." He asked for Army pilot Lt. Col. Madan K. C. (for "Khatri Chhetri"), who was related to a staff person at the U.S. Embassy.

At 6 a.m. on the 13th, the *Everest* team, Todd Burleson, Pete Athans, a group of Sherpas, and a number of others guided Beck Weathers down from Camp II, while nine Sherpas pulled Makalu Gau in a makeshift sled. They also carried the corpse of Chen Yu-Nan. The other Taiwanese were at Base Camp.

All were dreading the extremely difficult evacuation of Gau and Weathers through the Icefall. When they were partway down the Cwm, Base Camp radioed the evacuation team that the wind had died there, and that Lt. Col. K. C. would be attempting to land a French Squirrel helicopter above the Icefall. Few of the climbers were hopeful.

But then a miracle took place. While traversing a flat area above Camp I, the evacuation team stopped to watch a small, green object, insect-like and incongruous with the glacier, slowly circle above Base Camp. Then, as they stood in disbelief, the chopper clawed its way toward them with a distant, thwocking rattle.

Ed realized the team had nothing with which to mark a landing spot. "Araceli said, 'Wait—I have some red Kool-aid!' She opened her rucksack, pulled out her bottle of Kool-aid and tossed it to me."

Lt. Col. K. C. flew above and past Camp I, at 19,500 feet, and continued in the direction of Camp II before turning to descend and attempt a landing. He was alone, having left his copilot at Base Camp to make the aircraft lighter.

Ed poured out a small "X" on a flat, 30-meter-wide strip between two crevasses. David dropped to one knee to signal the pilot, and Ed tied a bandanna to an avalanche wand, for a wind sock. The pilot came in slowly and cautiously, clearly uncomfortable with the landing site. As the chopper passed over the

deep crevasse on one side of the landing area, it briefly lost the bouyant "ground effect", causing the tail to tilt backward. The protective wire below the tail nearly hit the edge of the crevasse just as K. C. touched the skids to the snow.

He quickly lifted off again.

"Oh no, that's it, he'll never land—his pants are full already!" Robert agonized. But K.C. gained altitude and came in this time from directly above. He never put the entire weight of the helicopter on the snow.

"There was a tailwind, which is the most difficult kind to deal with," K. C. calmly explained later. Gusts of wind at ground level can disrupt the lift and control of the blades, which were already compromised by the altitude. And, if the chopper were to settle into soft snow or a hidden crevasse, the tail could sit back and damage the rotor, causing the helicopter to become lodged.

"Merely getting stuck at Camp I could be fatal to the pilot," said Dr. David Shlim. "Unacclimatized, he would not live long at that altitude unless he were on oxygen constantly."

K. C. was wearing his oxygen mask, and his attention remained riveted on the controls. With one hand he signaled that he could take only one passenger.

"There was no way I could get into that helicopter and leave Makalu Gau there," Beck recalled. Makalu wasn't able to walk, so his evacuation through the Icefall would be especially hard. Makalu would go. The Sherpas quickly loaded him in behind the pilot. "At that point, my heart sank," Beck recalled.

The ship hovered upwards tentatively, nosed slowly forward, then dropped like a

Bundled aboard the rescue chopper at last, Beck Weathers gets his ticket to deliverance. Although risky, the flight enabled him to bypass the even greater perils associated with crossing the Khumbu Icefall.

rock out of sight down the Icefall. Lt. Col. K. C. set down at Base Camp and offloaded Makalu Gau.

An interminable thirty minutes passed. Beck looked toward Pete.

"Any guesses on whether he'll be back?"

"The breakfast buffet at the Yak and Yeti Hotel should be opening about the time you get there," Pete answered with a smile.

K. C. returned. "What a tremendous relief that we wouldn't have to carry Makalu or Beck through the Icefall," Ed recalled. He scratched his head and sighed. "It would have taken a day and a night and placed many people at significant risk. K. C. was a 'thank God' kind of guy. For me, the best part of the expedition was Beck's excited, joyful—and slightly incredulous —expression of deliverance." It was one of the highest helicopter rescues in history.

Landing again at Base Camp, K. C. picked up Makalu Gau and the co-pilot, and transported them all to Kathmandu. When they landed on the warm Kathmandu tarmac, Beck was in tears, and patted his bandaged hands on K. C.'s back. He told Madan how grateful he was and that he knew how much courage it took.

"I had always been told that I was given a brave heart," K. C. said later, "but until this rescue I never had the opportunity to find out if it was true."

RETURN TO
THE DEATH ZONE

*"You don't assault Everest. You sneak up on it,
and then get the hell outta there."*

—ED VIESTURS

The *Everest* team members descended through the Icefall once again, and arrived at Base Camp exhausted, frazzled, and shocked. Like the guides and clients who had returned the day before, some of them cried, finally able to begin releasing the anguish they felt over the tragic events on the mountain. ¶ Here, as they removed their crampons, they also learned that in addition to the five climbers lost on the south side, three Indian nationals also died below the summit on May 10, climbing on the north side of the mountain. ¶

Grisly reminder of past Everest failures, this partial cadaver lingered several years off-trail, at the base of Lhotse, before being investigated by climbers in 1996. The remains were consigned to a crevasse—a practice common to high-altitude climbing.

A MEMORIAL SERVICE

On the afternoon of May 14, two days after most climbers had returned to Base Camp, more than 60 people congregated at an informal memorial service for the lost climbers, held at the stone lhap-so constructed by Scott Fischer's expedition. The weather was cold and gray. "Only a few days earlier the mood had been one of excited optimism, and now suddenly we were attending a wake,"

Brad Ohlund observed. Like most others, he was shocked by the deaths and could only partly comprehend what had happened.

Neal Beidleman, Fischer's friend and assistant guide, opened the memorial fighting back tears and paused frequently to weep. Wongchu directed the Sherpas, who ignited juniper branches at the base of the small rock altar. As fragrant smoke spiraled toward the mountain, the Sherpas chanted prayers.

Offerings of biscuits and candy bars were passed around as communion. Rob Hall's sirdar Ang Tsering said, "When I see our good friends die while climbing, this job isn't much good."

Many climbers' hands and feet were bandaged, and they looked like war-zone survivors. Few people spoke until team members, mostly from the Fischer and Hall groups, took turns stepping forward. Some recited poems. Many cried. Some invoked memories of past times, as if reluctant to let go of their friends. "As each victim was remembered, I looked up toward the Icefall and had a feeling that even at this late stage they might come trudging home, down through the seracs—especially Andy Harris, who had simply disappeared," Audrey Salkeld recalled.

Veteran climber Pete Schoening, who had heroically saved several team members on K2 in 1953, offered a moving tribute. Looking much like a grizzled Alaska pioneer in a faded green duvet with fur-trimmed hood, he celebrated Scott Fischer's vitality and ability to inspire others.

"The atmosphere was a bit ominous, and it felt early to be having a service for Scott and Rob," Pete Athans observed, "but a tribute was needed. The gathering of people from all the groups, most of whom had worked together quite effectively, was especially good."

Araceli Segarra offered a European perspective, noting that the memorial was perhaps too open and talkative for something as final as death. "Perhaps this is how it is done in America. They must feel great grief, yes, but we Europeans don't feel comfortable showing our pain to everyone."

"But wounds don't heal when grief and emotions are bottled inside," Paula said.

The memorial marked the last time all of the teams would be together. With the departure of Hall's and Fischer's groups the next morning, the climbers who stayed felt more and more alone with their thoughts and the mountain. As one of them walked sadly over the rubble fields, through the abandoned tent sites, he said he could picture the former hustle and activity, as if the teams were still there.

Looking "like war-zone survivors," exhausted and pensive climbers hold a memorial service for their fallen comrades at a Base Camp lhap-so, four days after the tragedy.

REFLECTION AND ASSESSMENT

After the ceremony, the *Everest* team retired to the kitchen tent and sat in virtual silence. Over the next few days, as they ate, relaxed, and slept in a relative sea of oxygen, they were able to think more clearly about the recent events and to speak about them. Audrey felt she could hear the words that Colonel Edward F. Norton had written after Mallory and Irvine were lost near the summit on the north side in 1924:

"We were a sad little party; from the first we accepted the loss of our comrades…and there was never any tendency to a morbid harping on the irrevocable. But the tragedy was very near; our friends' vacant tents and vacant places at table were a constant reminder to us of what the atmosphere of the camp would have been had things gone differently."

News of the tragedy broke out in the press

and across the Internet. *Newsweek* was preparing a cover story. Liesl Clark of NOVA Online reported that the Web site covering the climb received more than 100,000 "hits" each day during the week that followed the tragedy. "The press accounts, television, and other media reminded us of what happened, which helped us confront our fears, deal with our grief, and heal," Paula said.

Along with friends from other expeditions, the climbers assessed the tragedy, not intending to assign blame to individuals—if this were even possible—but to make sense of the catastrophe and to learn from the mistakes that were made. The tragedy, they agreed, resulted from the confluence of bad luck and poor judgment. The guides and clients together had cut their safety margins too thin.

The month of April and the early part of May had seen only one or two full days of good weather high on the mountain. Robert noted that some survivors referred to the weather on the 10th as an unexpected or freak storm. But he and others were not sure the storm was all that unusual. Everest climbers should always be aware of and prepared for such conditions, he stressed.

"The events of May 10 were not an accident, nor an act of God," Jim Williams, a guide, said flatly. "They were the end result of people who were making decisions about how and whether to proceed. Unfortunately, not all the guides were really given the leadership or operating protocol for dealing with the various situations that arose on the mountain. The organization was all very loose."

Hall and Fischer had developed a plan to fix ropes high on the mountain well ahead of time. This didn't happen, and led to a situation where difficult decisions needed to be made very quickly. Hall's client Doug Hansen faltered on his descent from the summit, and Hall was faced with an agonizing decision. "Rob chose to stay with his client until he died," Ed Viesturs said. "Had it been me, right or wrong, I might have descended when I realized there was nothing I could do for Doug, even though he was alive. But Rob was a dedicated man."

"I don't believe you can hold any individual responsible for others' deaths. People died because of their own personal decisions," said Lou Kasischke, who was climbing with Rob Hall. Just below the South Summit, when one of the climbing Sherpas told him it was still two hours to the summit, Kasischke realized it was time to turn around. Two others on Hall's team also turned around near the South Summit, at 11:30 am. The views around them were already blocked by clouds, and they sensed that the weather was changing.

"We felt, because of the bottleneck delay and our 1 p.m. turn-around time, that it was too late to go to the summit, and would therefore be too risky coming back down," Kasischke said. "I didn't rescue anyone, and on summit day did nothing I can take pride in— except that at the critical moment I exercised the personal responsibility that each of us had and made a decision to turn around."

Since the tragedy, the *Everest* team and other Mount Everest climbers have reviewed and dissected the dynamics behind the events of May 10, and the decisions made. No single

RETURN TO THE DEATH ZONE

decision made on the mountain was necessarily wrong, they concurred. But if one or two decisions, out of many, hadn't been made, or if no storm had come, the outcome may have been very different. In the end, lives were lost as a result of compounding factors:

Weakness in numbers

When multiple teams ascend together, they benefit from extra manpower when breaking trail and fixing ropes. The large number of climbers, however, contributes to bottlenecks and delays at places such as the Hillary Step. David Breashears identified a "tagalong" factor at work, too: When large, experienced groups such as Hall's and Fischer's make their push for the top, less experienced parties and individuals imagine greater strength in numbers— and perhaps assume that someone will be available to help if something goes wrong. But climbers don't have a lot of time or reserve strength to wait for a slow climber or to rescue an incapacitated person at extreme altitudes.

"After a point, larger numbers only reduce a climber's chances of success," David explained. "Out of a handful of people, some are bound to have trouble, and when they do, everyone is placed at risk because any high-altitude rescue endangers the safety of the rescuer. When someone is in a life-threatening situation, I'm going to abandon whatever I'm doing to help, and I hope that others will as well."

Dr. Tom Hornbein agrees. "I've written off going to the popular West Buttress of Denali because I'd feel obligated to help anyone injured, and I don't want

to get killed helping a stranger who possibly has no business being there in the first place. The events that happened last May will eventually repeat themselves in some form."

"It's a feature of Western society to expect that if you're in a car crash, an ambulance will pull up and save you," added Todd Burleson. "That's not going to happen in the mountains."

Pete Athans and Todd were the only climbers to reach the South Col during the rescue period, fighting gale-force winds to get there. Like others who observed the scene on the South Col, they came back humbled. They downplayed their own heroic and life-saving efforts, and pointed out that the only true rescue was made by the Sherpas who retrieved Makalu Gau from below the Southeast Ridge. "Anytime we try to confront forces of nature so powerful and sublime as those we saw this year on Everest," Pete said, "we realize exactly how helpless we can be, how insignificant are our actions."

Inexperience

"There were many who seemed confident and brave at Base Camp, like elegant young men and women, yet they took several hours to climb only a short distance," Robert Schauer noticed, "and some of their climbing techniques were awkward and inefficient." One climber pointed out that some of the guides, too, had never been above the South Col. "How could such a person know from experience the types of difficult decisions that need to be made, especially at a place like the South Summit?"

FOLLOWING PAGES:
Softened by alpenglow, the stark beauty of the Himalaya rules this view from near Base Camp, embodying both the challenge and the reward of an Everest trek.

Turnaround time

Even before arriving at Base Camp, all climbers were aware of the importance of descending from the summit in time to reach the South Col before nightfall. The importance of a strict turnaround had been stressed to the guided clients, but on the mountain, these plans were disregarded by all but a few clients and guides.

Ken Kamler had been high on Everest the year before, and sensed that Hall, Hansen, and the New Zealand team had stayed high on the mountain too long that year, too. Some team members returned to Camp IV with frostbite.

Lou Kasischke remembered that before he reached the Southeast Ridge in the early morning of May 10, Doug Hansen stepped out of the trail of ascending climbers. When Kasischke passed him, Hansen said that he was tired and was planning to turn around. Before actually descending, Hansen changed his mind and continued upward.

Communication

"This tragedy," said Pete Athans, "has certainly reconfirmed the need for a *non-negotiable* turnaround time, for good support on the mountain, for good reserves of bottled oxygen, and for good communication."

Among those in Fischer's summit party, only Fischer and Lobsang Jangbu, his sirdar, carried radios. Mike Groom, a guide for Hall, had a radio that didn't work. Beck Weathers later pointed out that if the party of his team members that turned around and descended had a radio, Hall would likely have told him to join them. And if at least one of the lost climbers arriving on the South Col had a radio, they might have found their way to

camp, or their rescue may have been expedited. Tom Hornbein suggests that radio communication, not only between leaders and their guides but especially between leaders of the two teams, would have enabled the leaders to make a collective decision and possibly turn both parties around. With no ability to discuss all the factors at work on the mountain, it is more difficult for a leader to make a unilateral decision to turn around while the other group is still heading toward the summit.

Guiding without oxygen

Most guides, including Ed Viesturs, use supplemental oxygen when guiding at extreme altitudes, knowing that their ability to react to events is improved by the additional mental and physical energy that oxygen provides. On May 10, Scott Fischer's sirdar, Lobsang Jangbu, was performing many of the duties of a guide, and he summited without bottled oxygen. Anatoli Boukreev, guiding for Fischer, also climbed without it. During his team's summit push, Boukreev spent little time near the clients and, after summiting, descended to the South Col ahead of them. He was in his tent when they became stranded. Although he is rightfully credited for going out later and saving three of them, his decision to climb without oxygen and leave his clients to fend for themselves has been criticized by many veteran climbers.

Boukreev argues that if his bottled oxygen runs out, the shock of losing that supplementary oxygen makes him worse off than if he were breathing only ambient air all along. Dr. Hornbein doesn't question what climbers say they experience, but points out that this

RETURN TO THE DEATH ZONE

argument has no scientific basis. The blunting of judgment that occurs at high altitudes affects veteran guides as easily as it does anyone else. Climbing Everest without oxygen can be rationalized only in terms of personal achievement, not as a safety measure.

Boukreev also explained that he waited on the summit for an hour before becoming cold, a potentially dangerous situation. He thought that by descending in advance of his party he could bring oxygen up to those arriving late on the South Col, though weather prevented this.

Ambition

An obsession with attaining the summit played an overarching role in the tragedy. Many agree that Hall and Fischer were under pressure to succeed: clients had paid them substantial amounts to help get them to the top. It might be assumed this would manifest itself in overt pressure from clients, but Yasuko Namba and Doug Hansen were not the type to challenge Hall's advice. Rather, guides placed high demands on themselves to build a record of success in the lucrative guiding business. All climbers were aware that retreating when halfway to the summit would likely necessitate a return all the way to Base Camp, with little hope of a second try.

The press and other media may also have played a role, especially with *Outside* magazine journalist Jon Krakauer and NBC correspondent Sandy Pittman among the clients. With the world watching, and with a corresponding pressure to succeed, it would have been easy for Hall and Fischer to get swept up in a spirit of friendly competition. "It's hard to imagine that it didn't affect Rob to see all of Scott's clients

approach and then reach the summit ahead of his, especially after four of Rob's team turned around early," Williams said. The year before, none of Hall's clients had reached the top.

The morning of the summit push, Fischer had left his tent an hour behind the others, climbed slowly, and never managed to catch up with his clients. He may have been suffering altitude sickness—or he may have been just plain sick—and in hindsight should have decided to turn around earlier. "Scott was as strong as an ox, but something happened to him," Todd Burleson said. "It's scary. He has climbed Everest before without oxygen, and now, with oxygen, he didn't make it back down. This tells me that even as professional guides we're susceptible, that we have to watch ourselves closely, we have to be prepared to turn back."

"Scott was charismatic, good-hearted, and thought well of people—he had a positive attitude that you might call innocently enthusiastic," David said. "He was as much a cheerleader as an organizer. He may not have realized what utter chaos can occur on an expedition, that situations can arise that seriously test the limits of our control.

"Guides and climbers are ambitious, but ambition does not make you stronger," he continued. "It can get you in situations where you shouldn't be. I encountered guides who cheerfully stated, 'Everyone's doing great. We're all going to make it to the top.' And I thought, well, they've reached Camp II. Let's see how they do at Camp III on the Lhotse Face, and then Camp IV. There's a human sort of optimism there, but it borders on the cavalier. On the other hand, it may sometimes be good leadership. It's a hard thing to define."

"Unbridled ambition can kill you," Lou Kasischke, 53, says frankly. "And it almost killed me. I wish I had never gone to Everest. Given my family responsibilities and my age, I can't reconcile taking those extreme risks."

SHERPA ASSESSMENT OF THE TRAGEDY

The climbing Sherpas believe that tragic events are not always simple, and that factors such as luck, astrological alignment, and ripening of accumulated karma—along with judgment— play major roles.

Jamling emphasizes that the goddess Miyolangsangma can be defiled by people abusing the mountain—polluting it with garbage or attempting to climb it without proper respect. "The goddess can respond by causing the weather to change, by triggering avalanches or accidents, or by blocking the path down the mountain. I believe this partly explains what happened." Jamling and other Sherpas agree that foreigners are "excused" by the mountain divinities—but only to the extent that they are ignorant of these processes.

Can one really prepare for a wrathful Everest? It is relatively easy to train for Everest physically and to plan for it logistically.

Triumphant without bottled oxygen or climbing partners, celebrated alpinist Reinhold Messner soloed Everest in 1980. The tripod, left five years earlier by a victorious Chinese team, has since disappeared.

But mental preparation, Sherpas recognize, means developing mindfulness and right motivation. A goal can never be reached through force, former Tengboche monk Phurba Sonam points out, or by aspiration and ambition alone. But if the nature of the motivation is pure, stemming from a compassionate desire to help others, the goal will almost always be reached eventually.

DETERMINATION OF DEATH

Those who set out to rescue the victims and survivors exhibited a large measure of that compassion, and the Sherpas believe the rescuers will earn *sonam,* or merit, for their actions. But victims and rescuers alike have great difficulty in responding "normally" at extreme altitudes, because hypoxia and exhaustion rob them of their judgment.

"I don't think the doctor that looked at me should be faulted for declaring me dead, or close to death," Beck Weathers later said with calm sincerity. "It may not have been the best diagnosis, but we all make mistakes at times, even at sea level. He's an excellent doctor and a great guy, and he was the one climber of many on the South Col who ventured out during a period when there wasn't any great stampede of the cavalry to look for survivors."

Many wonder how Beck could have been identified as dead, when he later stood up and walked away. "Weathers's resurrection complicates the already prickly process of judging a person's medical condition in a harsh environment that only grudgingly gives back life," said Dr. David Shlim.

Unless Weathers's resurrection is explained as

a miracle—and Beck himself doesn't discount an element of the miraculous in his survival and rescue—then surviving as he did must be considered possible for anyone. "The beautiful thing about Beck's recovery is that not all of it can be explained," said Dr. Charles Houston.

What are the implications of Beck's survival in terms of the efforts that should be taken to save others stranded in extreme settings? At present, climbers, medical people, helicopter pilots, and others apply their years of experience and the best of their abilities to rescue the injured and stranded. Should the knowledge that someone may—but more likely may not—be still alive in an isolated, dangerous location inspire already committed rescuers to endanger themselves to an even greater degree?

"The fact that Beck was left for dead and survived is going to haunt rescue decisions for years to come," Shlim said. "But ultimately, those who can be pulled to safety will be, while those who appear lifeless or for whom not enough resources are available, will be left behind."

Yasuko Namba, for example. She died on the South Col near the spot where Beck Weathers lay before he arose. Perhaps Namba—and others in the past—may have had a longer window of survivability than people assumed. She may have been alive on the morning of the 11th, and she was lying only 400 yards across relatively flat terrain from the tents of Camp IV.

The confusion over who should have been caring for Beck Weathers was also unfortunate. Having stunned everyone by surviving the storm and a night alone on the South Col, Beck was then inadvertently neglected in his

tent in Camp IV and left to either suffer or die alone. Remarkably, he defied the odds against his surviving a second excruciating night, and was led down to Camp II the next morning.

ALTITUDE AND JUDGMENT

But the conditions on the South Col are difficult to imagine. Given what the people who went out to check on Weathers and Namba knew, and in view of the frightful conditions, they likely made the most appropriate decision. "Easy though it may be to second-guess them," Charles Houston says, "one cannot fault them for what they decided in their brain-numbed condition."

Diminished awareness and limited judgment

Dehydrated and exhausted after his solo sprint, Messner recovers in Advanced Base Camp, some 21,300 feet up on the Tibetan side of the mountain—along a route pioneered more than 50 years earlier by Mallory and Irvine.

is thought to have contributed to many of the "accidental" deaths on Everest. At 26,000 feet, people simply don't think the way they do at sea level. One of the first symptoms of hypoxia is loss of some mental faculties, especially judgment.

Hallucinations are common under hypoxic stress. "In 1933, Frank Smythe fed a bite of mint cake to an unseen companion, and saw strange flying objects over the North Ridge which his friends jocularly called 'Frank's pulsating teapots,'" Audrey observed. "And Reinhold Messner's companion Peter Habeler had an 'out of body' experience when, floating above his own shoulder, he watched himself climbing the upper slopes."

A TOUGH DECISION

"After the tragedy, I felt very mortal and very humble," Breashears recalled. "The mountain ceased to be a source of joy for me. Suddenly the wind seemed louder, the cold colder, my legs weaker, and the mountain higher. I thought, 'Wow, Breashears, by putting yourself and the team at this kind of risk, you've bitten off more than you can chew this time.' But the death of a close friend on another expedition tends to impact you less than when one of your own party dies. Our group was still intact. I wanted to ruminate for a few days over whether to carry on with our attempt, and to make a decision after talking with the team. Thankfully, there was no pressure on us from MacGillivray Freeman Films to continue."

Ultimately, whether or not to return to the mountain would require a personal decision from each member. Ed didn't want to leave with a pall of gloom hanging over the mountain. He wanted to remind himself and demonstrate to others that Everest can be climbed safely, and even be enjoyable and rewarding. "Everest isn't necessarily a death sentence or some sort of penance," he said.

Jamling's situation was more complicated. His wife, Soyang, and other relatives were opposed to him returning to the mountain; it had been hard enough for him to convince Soyang in the first place. Jamling asked her to again consult Geshé Rimpoche, their family guru and adviser, and request a new *mo,* a div-

Ever present danger: The awesomely perilous Khumbu Icefall poses the first and final hazard for every climber on the South Col route. Since 1921 it has claimed at least 19 lives.

ination. Soyang agreed that if the mo was favorable, she would relent.

In Kathmandu she told Rimpoche about the tragedy and the weather, and Jamling's desire. He again consulted the beads, then gave her his answer.

Soyang called Jamling and relayed Rimpoche's finding, and from the lightness of her voice Jamling immediately knew the response: "'Go! Go up! The circumstances haven't changed for you.'"

If the mo had been unfavorable, Jamling said that he would have respected her wish and abandoned a second attempt. He admitted that, were his family not involved, he might have proceeded with the climb anyway, but only after taking additional precautions, by commissioning further pujas, and by making offerings and prayers.

"Doing pujas and obtaining blessings from lamas is important and can be helpful," Jamling stressed, "but the critical element in climbing Everest, for Sherpas as well as foreigners, is having a strong *lungta,* wind energy, and right motivation. These determine our destiny on the mountain and are ultimately within our control. We must exercise good judgment, self-restraint, and respect for the power of nature."

But the period of bad luck wasn't completly over. Nawang Dorje, the Sherpa with Scott's team who had contracted what may have been a complicated case of pulmonary edema, was evacuated to Kathmandu from the

Pheriche Aid Post, his condition deteriorating.

The team's climbing Sherpas sat about Base Camp, demoralized and unenthusiastic about returning to the mountain. Some of the Sherpas stopped in to see Pheriche Aid Post doctor Jim Litch with vague complaints, though he felt they visited more to get a prescription of hope and confidence. Ever practical, other Sherpas were concerned that the deaths might result in fewer Everest expe-

dition jobs. But guides and Sherpas would soon find that the events of the spring season appeared to have no such effect.

Araceli wanted one more shot, having put so much time and energy into the expedition. But second thoughts crowded in. "I don't want to climb on a route with dead people on it," she said, crying.

"This has been a big shock for me, I've never seen such serious frostbite before, and I've

JUTTING INTO THE JET STREAM

Each summer, tropical heat over the Indian Ocean creates a large mass of humid air that is drawn across India, toward the Himalaya, by the convection currents rising over the massive, 5 kilometer-high Tibetan plateau. When this water-saturated wind, associated with the summer southwest monsoon, strikes the Himalaya, it rises and cools. The cooling causes the moisture to condense in the form of heavy rain showers—and snow at higher elevations—mainly on the south side of the range.

But the meteorological effects of the Himalaya and the Tibetan plateau extend beyond South Asia: The range and plateau are known to effect on the planet's jet streams, and thereby influence global weather patterns.

Jet streams are masses of fast moving air that are found between six and nine miles above the earth's surface. Typically, they are thousands of kilometers long, a few hundred kilometers wide, a few kilometers thick, and travel west to east. Jet streams arise where there is a strong temperature gradient, and are therefore stronger during the winter months, when temperature contrasts are greater than during summer.

The Subtropical Jet flows along the front range of the Himalaya; some feel that many, and possibly all, northern hemisphere weather sequences originate on the Tibetan plateau.

There's an extraordinary "shear" right at the base of the Subtropical Jet, typically at about 27,000 feet.

"One of Everest's inherent dangers is that this shear level is quite close to the summit," meteorologist Bob Rice noted, "such that the winds on the summit can go from 25 to 150 knots almost instantly." It is thought that some of the climbers lost on Everest in the spring of 1997 were literally blown off the north ridge when the jet stream descended onto them.

"The core of the jet stream is like a ribbon, moving in all three dimensions," Rice explained.

"The jet's horizontal undulations tend to be broad and are relatively predictable, but the up and down movements occur in waves, like a snapped rope, within only a few thousand vertical feet, and are unpredictable." It is therefore best to climb the mountain when the jet stream has moved north and is known to be out of the way.

The *Everest* team were pinning their hopes on British meteorologist Martin Harris, whose forecasts of jet stream movement had so far been remarkably accurate. For additional input, Liesl Clark contacted Bob Rice's "Weather Window," while Roger Bilham called weather scientists of the National Climate and Atmospheric Research Center (NCAR) in Boulder. All were scrutinizing the satellite images that are issued periodically by the National Weather Service, and which are supplemented by supercomputer predictions on their development trends. Shortly before May 23, the images showed that the jet stream was moving northward.

Forecasting the weather and the movement of the jet stream is "a bit like reading tea leaves," Roger pointed out. But meteorologists with a computer and access to satellite weather data—have made forecasting a valuable, and potentially life saving, tool for climbing Everest.

Distinctively shaped lenticular clouds crown Everest and Lhotse, signalling a strong jet stream, high winds, and—usually—several days of clear weather.

thought a lot about the dead and the survivors," Sumiyo added. "I didn't know Yasuko well; she had climbed the highest peak on every continent, and Everest was her final summit. But this doesn't make me afraid for our team, because I trust each of them. We are strong, experienced, and ready."

"I never thought about *not* going back up," Robert said. "We had a job to finish, and actually felt that the window of calm weather hadn't yet come. The other teams that remained suspected this, too."

It was unanimous: They would try to get as high as safely possible.

Paula Viesturs was unprepared for the decision. She was tired and stressed, and became angry at the first news of it. Ed was certain it would be harder on Paula than on him. The tension at Base Camp had been unrelenting, and she decided to take a walk for a few days, knowing it would be that long before Ed and the team would be high on the mountain again. She hiked down to Tengboche with the departing New Zealanders. She wanted to clear her head, to see the flowers, to decompress.

"The mountain is a place of astonishing beauty," David said expansively. "Despite all the work, the tragedies, and the setbacks, we wouldn't be going up again if there wasn't also a lot of joy in it. Seeing the younger climbers' drive and enthusiasm also helps Ed and me. But none of us are afraid to come home without having made it to the summit."

As Base Camp manager on Everest, Paula Viesturs oversaw the ordering of food and preparation of meals for the entire Everest *Film Expedition crew.*

"As experienced climbers, we have all lost good friends to the mountains over the years," Jim Litch reflected. "In spite of this, we continue to climb. There must be an element of denial that it could ever happen to us; otherwise, how could we knowingly continue to put ourselves at risk?"

RESTOCKING THE SOUTH COL

While members of most other teams were headed home, the *Everest* team was trying to find the strength, and the bottled oxygen, to head back up the mountain. During the rescue period they had volunteered their oxygen, as had all except the South African expedition. In order to film, a summit team of 11 to 13 people would be needed, with a budget of four or five bottles per person—a total of fifty bottles. Twenty-eight bottles had been expended on the South Col during the tragedy; restocking the South Col would be at least two days' work for the Sherpas.

Todd Burleson's team had oxygen, but they also planned to make another attempt. Guy Cotter offered to replenish most of the *Everest* team's oxygen supplies, because Hall's team had expended much of their oxygen on the South Col.

THE MONSOON APPROACHES, BUT THE WIND IS THERE ALREADY

The grumbling and shifting of the Khumbu Icefall was growing, and the route through it was becoming difficult to maintain. Mal

Duff's team, which had installed and repaired the route throughout the spring' had departed, but two Sherpas remained to work on it.

On May 17 the team again left Base Camp, and by late morning they topped the Icefall. Again, they plodded into the reflected heat of the Western Cwm and the blinding glare from the layer of smooth, melted snow ahead. Early that afternoon they arrived at Camp II, which had become like a second home. They would wait here for the wind to relinquish its grip on the mountain.

In their tents at night the team heard the ominous roar of wind on the mountain. "We needed calm weather soon, because our climbing permit ran out on the first of June," David said. "But our motivation might have run out first."

They were also thinking about the approaching monsoon, the moisture-laden winds from the south that build up in late May. David noted that the mornings in the second week of May were clear, as usual, but the clouds seemed to be forming just a bit earlier in the day. He was frustrated. "In 1985, Stein Aasheim from Norway climbed Everest, carrying a kite in his pack that he hoped to fly from the top—but there wasn't enough wind to get it up."

If the high winds continued, pushing for the summit would be difficult, and threading film through the camera would be impossible. At the least, they hoped to complete the mission of installing the weather station on the South Col, which Roger had prepared them for.

Testament to human frailty as well as Everest's ferocity, abandoned and wind-shredded tents litter the Camp III ice. Though these eventually will blow away, more inevitably will appear.

Base Camp was windy, too. Brad Ohlund estimated the wind speed at 40 miles per hour, and he worried that his tent, with him in it, would get blown, bouncing and tumbling, across the glacier. "For three days my radio inquiries as to how the climbers were doing were answered in the same way," he reported. "'We're cold and tired. We want to get this done and go home.'"

The fatigue of a long expedition had begun to set in.

Pete Athans, Todd Burleson, and Jim Williams had also returned to the mountain with two of their five clients. They settled in beside the *Everest* team at Camp II and waited five days before abandoning their attempt. Their clients didn't feel good about the mountain, and elected to return home.

"If we had reached the top, having to guide paying clients past the corpses of our friends would have brought little joy," Pete said. "And I have to admit that our decision to retreat was partly influenced by the international news media. With all the attention focused on the mountain, the press would crucify us if a client was harmed, and we could not justify the risk."

Along with the team, a Spaniard, a Frenchman and Göran Kropp, the lone Swede, decided to remain on the mountain. The South Africans also announced their plan to try for the summit again, which worried some of the climbers on other teams.

By May 20 the reports and satellite images showed that the weather for May 23 looked

fairly good. The weather was improving daily, and during the daytime it became hot inside the tents at Camp II. Sitting at 21,300 feet in shirt sleeves, Ed looked ahead to the climb. "We have to assume that not all of our team will reach the summit," he said. "Anything can happen: lack of desire, illness, or other variables that are out of our control. And especially when you've been on the mountain for eight or ten weeks, you're homesick, tired, and have lost weight—and

that's right when you reach the higher altitudes and the going gets *very* tough. The climbers who have patience, persistence, and reserves of motivation will be the ones who summit."

And reserves of faith, too. Before leaving Base Camp, Jamling had hiked over to Kala Pattar and unrolled another length of prayer flags. Now he strung the five-colored flags at Camp II, as he would at the higher camps. For several minutes he stood motionless, praying.

CHAPTER SEVEN

"CHOMOLUNGMA HAS LIT A LIGHT WITHIN ME"

"…the summit of Everest can deliver you from the prison of ambition."
—PETER BOARDMAN

On the night of May 20, the team finally received confirmation from meteorologist Martin Harris in England that the jet stream had moved to the north. A window of good weather was opening. ¶ David Breashears described the summit plan: "In the early morning of May 23—shortly after midnight of the 22nd—we will leave the South Col and head up. Four Sherpas will assist with the camera, two Sherpas will carry oxygen for the others, and two will cache additional bottles on the Southeast Ridge for the returning climbers. No Sherpa will carry more than 35 pounds." ¶ If all went according to plan, the summit team of 11 climbers and Sherpas would be on top by 10 or 11 a.m., and back on the South Col between 2 and 4 in the afternoon, giving them a safety margin of a few hours before darkness set in. "Keep your fingers crossed and

Early morning magic: Viewed from the Balcony at dawn en route to the summit, Everest's triangular shadow floats dreamlike in the distance, falling upon nothing more substantial than air.

say your mantras," David advised. "Now, we only need that good weather."

TO THE COL

The team then attended to important details such as taping insulating foam to the head of their ice axes. Gripping steel in minus 30-degree temperatures would conduct the cold right into their hands, increasing the risk of frostbite.

On May 21 the team set out for Camp III. When they arrived that afternoon, they crawled into the tents perched on the snow shelves they had dug into the Lhotse Face. Over the past few weeks, they had climbed to Camp III and descended to Base Camp four times.

The next morning they continued upward. At the Yellow Band, known to geologists as the Chomolungma Detachment, the world's highest fault, Ed reminded himself to stop on the way down to collect rock samples for geologist Kip Hodges. The team then crossed the Geneva Spur, and in early afternoon reached Camp IV on the South Col, at 26,000 feet.

Araceli was awed by Everest's proximity. "When I stepped onto the Col, I could hear the mountain murmur. It came out in full display—black, white, red, orange, and gold—

Besieged by thoughts of the tragedy, Jamling and the rest of his team set out from Camp III on May 22 to restart their own summit attempt.

FOLLOWING PAGES:

Ed Viesturs climbs through the Yellow Band as he heads for the South Col. On his return, he would gather geological samples here.

a magnificent pyramid of rock and ice. The wind had created a different landscape; the spent oxygen bottles, twisted aluminum and shredded nylon of abandoned tents made the Col look like a field of metallic snow flowers."

The team's storage tent was still standing; despite the high-tech materials used in their construction, tents are often destroyed by the Col's ferocious winds. Yasuko Namba's body was 400 yards away, but two other bodies lay in the

open nearby, though no one paid them much attention. Dealing with one's own life here at the lower edge of the Death Zone was more urgent.

David decided that one member of the team should remain on the South Col as a communications and safety officer. With regret he selected Sumiyo Tsuzuki, whose cracked ribs would have endangered her summit attempt.

"I was disappointed, but knew that my role in supporting the team on the South Col was important for assuring their safe return," she said later.

After setting up camp, the climbers dried out their clothes, mittens, and inner boots, which had become wet from perspiration. By 5 p.m. on May 22, they had finished melting snow and ice for their water bottles and had gagged down a few bites of food.

HANGING OUT IN THE DEATH ZONE

Doctors and Himalayan climbers recognize the Death Zone as a general term, but the dangerous effects of extreme altitudes on the body and mind are difficult to overstate. Above 26,000 feet, climbers generally don't want to eat, drink, put their boots on, or go outside—even the smallest amount of physical exertion requires an arduous effort. The body deteriorates rapidly at this altitude; sleep is barely possible and not restorative; and food is poorly absorbed. Humans cannot survive for extended periods above this height.

"Araceli put on her oxygen mask and fell asleep immediately!" Ed laughed. "I lay there for four hours remembering how hard it was to climb Everest without oxygen. Had I trained enough? Was I still mentally strong enough? I didn't want to let anyone down, including myself, and was anxious to get moving."

David and Robert Schauer had to work that night. Confined with their cooking gear and the film equipment in a three-man tent, Robert had to load four large IMAX film magazines while David checked the camera lenses and body—operations that were complicated by

Sherpa members of the Everest *team toil toward the Yellow Band burdened with loads of supplies needed for Camp IV and the final push to the summit.*

their cumbersome oxygen masks, which they frequently removed. Without bottled oxygen, however, they could function for only 15 minutes before their muscle reactions and movements slowed down.

At Base Camp, Paula was nervous. "I still have dreams and fears about Ed climbing. I've trained myself to be positive, and I practice envisioning Ed and the rest of the team standing on the summit and getting down safely. Each climb is a new situation, a new mountain, and I believe in our team's ability to make the right choices." Paula, Liz, Brad, and Wongchu decided to get some sleep, but it was a restless night.

THE CLIMB

At 9 p.m. on May 22, Ed arose, melted more snow for water, and forced down a pop tart, the last solid food he would consume for 48 hours. Knowing that he wouldn't eat on the mountain, he added some energy drink to two liters of water and placed the bottles in his pack.

It can take seven liters of water a day to remain fully hydrated at high altitudes, but climbers can't carry that much. Despite their dehydration, they experience little thirst and are too cold to think about drinking the half-frozen slush in their water bottles.

"I woke Araceli, called Paula on the radio, climbed out of the tent, and told David that I'd see him soon," Ed recalled. He left an hour before the others, at 11 p.m. Climbing without oxygen and breaking trail, he expected that the rest of the team would catch up with him.

The sky was black. Ed tried to remember from

his reconnaissance the day before how to skirt the crevasses of the steep ice bulge above Camp IV. "Probably by sheer luck, I hit the fixed rope at the bottom of the Triangular Face right on," he said. The rope led 800 feet up a snow gully, at a 30-to 40-degree angle, then onto a snow slope at 50 degrees. It was slightly windy, typical for early morning on the mountain. Paula's words over the radio echoed in Ed's thoughts: "Climb that mountain like you've never climbed it before."

He climbed methodically, kicking through the fresh snow. He looked back occasionally for the headlamps of the rest of the team. His breathing was rapid. Step... Pant... Step... Pant.... Each laborious step took five seconds, a pace that slowed as he gained altitude. Though the walking distance from the South Col to the summit is only a mile and a half, the climbers can cover the terrain at an average of only 12 feet per minute.

By now, David and Robert had broken ice into a pan and "brewed up." They put on their overboots and placed water containers and two oxygen bottles in their packs, along with still cameras, spare mittens, spare goggles, and a couple of candy bars. They checked the pressure on the oxygen bottles to make sure they were full, and rechecked the regulators to confirm that the

"World's highest junkyard," Barry Bishop said in 1963 of the Camp IV site on South Col, littered over decades with thousands of spent oxygen canisters and debris from previous teams. Cleanup efforts between 1994 and 1996 removed an estimated 750 cannisters; perhaps 1,500 still remain.

216

oxygen was flowing at a rate of two liters per minute. They clipped on their crampons and had a sip of tea before leaving. At midnight they joined Jamling and Araceli outside.

"While we lay in our tents, she didn't stop puffing," Araceli wrote of the wind high on the mountain. "But before we left the Col she became quiet, as if consenting to our climb."

The four went off into the darkness, their visible world defined by the reach of their headlamps; deep breathing and the crunching of crampons on the ice were the only sounds. Jamling felt the coolness of his mother's ivory rosary around his wrist. He chanted mantras while counting off the beads between his thumb and the crook of his forefinger.

The Sherpas also were ready to depart the Col, and they radioed Base Camp to check in. A Sherpa at Base Camp lit a juniper incense fire at the lhap-so. He would keep it burning until the entire team returned to Camp IV.

Sunrise on summit day found Ed Viesturs well up the Southeast Ridge, after departing Camp IV before midnight. This view, taken by him, includes the ridge's lower reaches (foreground) and the distant summit of Makalu.

TO THE SOUTHEAST RIDGE

Ed was making better time than he expected. "The knee-deep snow was a hassle, but it was exciting to be up there and I resolved to plow through it." After two hours of climbing, the beam of his headlamp illuminated a corpse, sitting upright. The face and upper body were covered with snow. It was Scott Fischer. "I found it hard to look at him," Ed recalled sadly. "I wanted to spend some time there, and decided that I would stop on the way down."

In the South Gully, just below the Southeast Ridge, Ed realized that the rest of the team wasn't catching up, so he dug a hole in the snow, sat down, and waited. Without bottled oxygen, he quickly became cold, and after waiting for 45 minutes he decided to move on. He waited again on the Balcony, at 27,600 feet, the first landmark of the Southeast Ridge. He was exhilarated to be so high on the mountain, witnessing the first glimmer on the horizon grow brighter and bathe the world in amber light.

The rest of the team plodded upward. "On summit day, I'm in a cocoon," David said, describing the final push. "I get into a mantra of rhythm, and every bit of focus and ambition goes into putting one foot in front of the other. I feel reduced to elemental thoughts, to a state of consciousness that emanates from the need to survive. It's not a creative mode. I become pragmatic and methodical, and monitor my body's functions and resources. With every step, I'm thinking about my pace, my breathing, my posture, the time frame for reaching the Southeast Ridge, and the weather—wondering whether, once we reach the South Summit, we will still be a group, with Robert and the camera somewhere nearby."

To stay warm, Ed had to keep climbing, and he continued to push through knee- and thigh-deep snow. David, Robert, Araceli, and Jamling waited at the Balcony for the IMAX camera, and were joined by Jesùs Martínez of Spain, Göran Kropp of Sweden, and two Sherpas. Thierry Renault of France, climbing on the South Africans' permit, was ahead of them. "As we

waited and watched the sun rise, I turned off my oxygen, but took a few lungs full every ten minutes to stay warm and active," Robert said.

Araceli was excited. "I felt euphoric when we reached the Southeast Ridge—partly because I was sharing that beautiful day with the people on our team. We had a good attitude and knew we were finally on our way to the top."

Jamling sat and observed the tiny flat spot below the Balcony—his father's and Sir Edmund Hillary's last camp before the summit. He remembered the story of his father's night there on the snow shelf barely large enough for a small tent. Jamling then gazed out over Khumbu, placid in the morning haze far below them. He felt he could hear the ritual chanting of monks, their pulse-like drumbeats and high pitched horns.

When Jangbu arrived at the Balcony with the camera, David and Robert set up a shot. They promptly realized they had lost their voices—not uncommon at altitude—from deeply breathing the dry, cold air through an oxygen mask, which dries the air even more. They communicated with each other in sign language.

From the Balcony, the team started on the 1,100-vertical-foot ascent to the South Summit. By the time they reached it, they had been moving continuously for nine hours.

THE SOUTH SUMMIT

"At 9 a.m. a voice croaked over the radio," Liz Cohen reported from Base Camp, "and we all

Lagging behind Ed by about two hours, other team members labor up the Southeast Ridge now in full daylight.

FOLLOWING PAGES:

Looking uphill from the same area where the previous photograph was taken, the team progresses along the Southeast Ridge.

jumped out of our seats. David and Ed were at the South Summit and the rest of the team trailed behind by at least one hour." The two waited there for some minutes before they became cold, then continued on. At that altitude, the simple act of breathing requires so much metabolic energy that it is not possible to stay warm for more than a short time—regardless of clothing.

Just beyond the South Summit, David and Ed came upon Rob Hall's body. "It was obvious that he had done all the right things," David said. "He had surrounded himself with extra oxygen bottles and removed his crampons to help keep his feet warm. He had applied his formidable willpower and mountaineering skill in a heroic attempt to survive. But Rob was a long way from help. Nobody could have survived in those conditions, and nobody could have saved him."

"Seeing Rob Hall's body was the hardest part of the ascent," Araceli said, "and he was right at a place where we most needed to concentrate—not a place to make a mistake."

They were now only 300 vertical feet from the summit, but had yet to surmount the treacherous Hillary Step. In *Tiger of the Snows,* Audrey recalled, Tenzing Norgay described the winding "snowy humped" ridge of the South Summit and the steep, rocky step—later termed the Hillary Step—some 40 feet high, that blocked his and Hillary's progress in 1953. Hillary discovered a vertical crack in the out-

cropping and was able to get into it and jam and wriggle his body upward. Tenzing followed.

As David moved ahead, he could see that the route up the Hillary Step had changed and was more awkward and time-consuming than in 1983 and 1985. Also, the route was laced with a confusing and entangling maze of old fixed ropes. Thierry Renault was ahead of David but moving slowly, spurring fears of another delay.

For Araceli, everything seemed to go wrong at the Hillary Step. Her nose started to bleed and she stopped, holding up climbers behind her. Then, when she placed her handkerchief inside her oxygen mask, her goggles fogged. "After I finally climbed the Hillary Step, I asked Jamling how far it was to the summit," she said. "Of course he hadn't been there either, but the summit ridge seemed to go on forever."

Having attained the South Summit, team members first traverse to the notorious Hillary Step, then proceed toward Summit Ridge—visible here—and the elusive true summit, still out of sight.

SUCCESS

At 10:55 a.m. Nepal time, May 23, Ed radioed Base Camp to say that he and David had gone as far as they could. "From where we are now, it's downhill in all directions," he announced from the summit. The Sherpas heard cheers from the dining tent and came in to join Liz, Paula, Brad, and Wongchu.

Jangbu caught up with Ed and David on top, and the three of them waited 20 minutes for the camera. Climbing sirdar Lhakpa Dorje was also moving slowly, having decided to climb without oxygen. Again, Ed was getting cold and had to get going—which meant descend. He passed Jamling and Araceli as they approached the top,

and gave them each a hug of congratulations.

Around 11:35 a.m., Jamling, Araceli, Robert, and Sherpas Lhakpa Dorje, Thilen Dorje, and Muktu Lhakpa reached the top. Araceli took out the Catalan flag, while Jamling and David took photographs of her. Over the radio she was connected to a reporter with Catalan television, and she declared that she was hungry for more chocolate.

Jamling may have been the most ecstatic. "The moment I reached the summit I felt a rush of excitement. This was where my father had stood 43 years ago. I hugged David and thanked him, because he had given me the opportunity to climb Chomolungma. I cried a bit out of joy, and as I looked around I put my hands together and said *thu chi-chay*— thank you—to Chomolungma. Then I prayed.

"I prayed that my father might be able to watch and be proud, and I prayed also for our safe descent."

Alternately facing in the cardinal directions, Jamling cast small handfuls of blessed rice into the air. He then unfurled the long prayer flag, and tied it to the cluster of other flags and katas adorning the metal survey stake anchored in the summit snow by an Italian expedition.

Jamling then struck his father's summit pose for the camera, not realizing that his stance was the reverse of that now famous image of Tenzing. He was connected by radio to his wife, Soyang, in Kathmandu, who was nearly breathless. "Now, no more!" she exclaimed and reminded him to be cautious coming down.

He placed the photographs of his mother, father, and His Holiness the Dalai Lama on the summit, along with the packet of blessed relics from high Tibetan lamas. Next to the photographs, Jamling left the rattle from his daughter.

When the IMAX camera arrived, David began filming. He kept his oxygen masks on almost constantly, in order to concentrate better, and shot one 90-second roll of film. "Having already climbed Everest, my main goal was to get the camera on the monopod and fire it up," Robert said. "When I heard the sound of the camera operating smoothly, I was thrilled, for this was an important reason we came to Everest."

"We accomplished something historic, and it was a wonderful moment," David said. "But I was also concerned that everyone get down as safely as they got up."

Jesùs Martínez reached the summit around 12:35. With him was Ang Rita, who had now climbed Everest a record ten times. Accompanying them was Thierry Renault, climbing with two Sherpas, and Göran Kropp, who was very tired. It was Kropp's third attempt that season.

A DIFFICULT DESCENT

Ed was moving downhill quickly. "At the South Summit, I sat down with Rob and just talked. He was on his side, and his left glove was off." Hall's wife, Jan, had asked Ed to take photographs of Hall's body. Viewing those photos

An exuberant Jamling Norgay strikes a top-of-Everest pose much like his father's, made 43 years earlier. His ice ax flies flags of India, Nepal, Tibet, U.S.A., and the U.N. Nearby, blessing scarves and an array of prisms, left by an Italian team for survey purposes, crown the summit.

later, Ed and David saw something that Ed hadn't noticed when he took the shot: a piece of material sticking out of the snow a short distance from Hall that looked as if it could have been covering a knee or an elbow. It may have been Doug Hansen, who was assumed to be near Hall, attached to the fixed rope.

Lower down, at 27,000 feet, Ed again sat down, and he asked Scott Fischer about what happened. David, too, was affected strongly.

"Scott was in a very lonely place, and it was sad to see him there. In another hour he'd have been in camp."

Neither Ed nor David was able to move Hall's or Fischer's body away from the route. "Rob's wife, Jan, and Scott's wife, Jean, had asked me to try to retrieve some remembrances from them," Ed recounted. "Rob wore a watch, and Scott wore his wedding ring on a thong necklace. But it was too immediate and too personal for me; I couldn't disturb them. I had always thought I'd grow old with them, and assumed that if they got in a situation like this one, they'd come through it alive. I'd never had a close friend die anywhere, let alone in the mountains, and here were two good friends—dead. Their faces were covered with snow—thankfully, because I wanted to remember them as I had known them. Alive."

On the descent, Araceli became lethargic and eventually sat down, unaware that her oxygen had run out above the Hillary Step. "It felt awful and dangerous," she said. Robert

ED VIESTURS, HVR, AND VO₂ MAX

One of the joys of climbing with Ed Viesturs is that you know you're in the presence of a superior being," said David Breashears, a powerful climber himself . "Ed has a rare ability to operate exceedingly well at high elevations."

"Ed chose his parents well," explained Dr. Robert "Brownie" Schoene, Professor of Medicine at the University of Washington. "Much of his ability is due to his genetics. But anyone who has climbed with Ed will tell you that he simply moves well. He's smart and careful, but bold enough to do great climbs."

Since the late 1970s, Dr. Schoene, an accomplished mountaineer, has studied the performance capabilities of endurance athletes and high-altitude climbers, and was a member of the 1981 American Medical Research Expedition on Everest. Before Ed departed Seattle for the Everest Film Expedition, Schoene tested him at the University of Washington's Pulmonary Function and Exercise Laboratory. Predictably, Ed scored high on every test.

To determine his response to hypoxia, Ed relaxed as he breathed into a mouthpiece that measured the quantity of air he exhaled. Over ten minutes, the amount of oxygen supplied to him was decreased, causing hypoxemia, or a low level of oxygen in the bloodstream—which in turn stimulated

his breathing. Ed's "hypoxic ventilatory response," or HVR, ranked high, meaning that the volume of air he breathed increased substantially as the oxygen supply was decreased.

"At moderate altitudes of 10,000 to 14,000 feet, where the available oxygen is only 60 to 70 percent that of sea level, a brisk HVR minimizes the susceptibility to some high altitude illnesses, and may improve performance," Schoene said. "But above 20,000 feet, where there is very little excess oxygen, adequate breathing is essential to insure that enough oxygen gets from the lungs to the blood, and into the tissues."

Schoene found that, in contrast to elite mountaineers, middle- and long-distance athletes tend to have a blunted, or low, HVR. This actually benefits them: There's a surfeit of oxygen at low altitudes, and it's more economical for them to breathe less, because breathing takes work and expends energy.

On the summit of Everest, about 30 percent of a climber's oxygen intake goes to the physical activity of breathing. "This means that 30 percent of the climber's energy is spent simply on survival," Schoene said. "At sea level, a normal person at an exhaustive level of exercise spends only 7 percent of his energy on breathing. In addition, breathing at altitude steals a lot of blood and oxygen from other organs—

including the brain, which leads to diminished cerebral functions."

Aerobic capacity, or "VO₂ max", by comparison, is the maximum volume of oxygen an individual can absorb at the end of exhaustive exercise, and is a marker of aerobic fitness. Schoene and others have determined that low-altitude endurance athletes have a high VO₂ max. "People then assume that elite extreme-altitude climbers must also have huge VO₂ max levels, because as one ascends, the ease of oxygen consumption falls," Schoene pointed out. "But some climbers defy this, particularly Reinhold Messner, the first person to summit Everest without supplementary oxygen. Messner's VO₂ max is higher than the norm, but is not exceptional. The relationship between VO₂ max and climbing ability is somewhat of a mystery."

Ed's VO₂ max tested very high, and he is now a spokesman for a company that produces a high energy candy bar called, appropriately, the "VO₂ Max."

Ed also has a high anaerobic threshold. This threshold is the point during a progressive exercise test beyond which one can't continue for more than a few minutes. Above this level, one develops lactic acidosis, or too much acid build-up in the muscles. Anaerobic threshold is expressed as a percentage of VO₂ max, and the average level is

about 55 percent. The anaerobic thresholds of elite athletes—especially those in endurance sports—are in the 90 percent range, which is comparable to the pronghorn antelope. Ed operates at 87 percent of his VO₂ max, meaning that he can exert comfortably for extended periods at a high percentage (80 to 85 percent) of his maximum energy output. He's aerobically fit and can function at a high percentage of his maximum (in endurance sports, anaerobic threshold is a better performance indicator than HVR). As with HVR and VO₂ max, training can raise one's anaerobic threshold, but not by much.

But other factors also enable Ed and other successful extreme-altitude climbers to move well in the mountains, such as an efficient biomechanical make-up. But following the tragic deaths of May 1996, many people have brought up the idea of testing climbers—especially guided clients—before recommending they climb Everest. "I don't foresee that the test we do will become everyday standards to determine whether or not people should climb," Schoene concludes. "Climbing successfully to extreme altitudes, and returning, requires a mysterious mix of characteristics that scientists can only partly define. I simply tell most people to listen to their bodies."

had to prod her into standing up and moving.

Jamling was at their cache of reserve oxygen bottles at the South Summit, and he put a new bottle in her pack. "'Wow! Oxygen again! Let's go!' was how I felt," Araceli said. "Jamling and I sped—*vroooom*—down the mountain. When we arrived at the South Col, three hours later, Jamling told me he had set the regulator at three liters per minute flow. I said 'What?—so that's why we were going so fast!'"

"The South Col no longer seemed a cold and lifeless place, and I was happy to arrive there," Araceli wrote in her journal. "The wind had stopped, and I turned and looked at the upper mountain. She stood elegant and powerful; an air of kindness surrounded her, or maybe it was my own sense of peacefulness at having stepped safely onto her icy skin." Araceli and Sumiyo hugged, and Sumiyo gave her tea. The team then collapsed in sleep.

David was relieved. "For me, the day wasn't joyful until the moment all of us were on the South Col and in our sleeping bags. We were rewarded for our patience and perseverance, and were graced with two days of fine weather. It's been a hard expedition for me. Standing on top was an anxious moment because of those we know who stood there less than two weeks before us, but didn't make it down. I'm ready to go home and relax."

"Ed did a great job breaking trail," Robert said, "and it's amazing that he did it without oxygen. All night we saw only a tiny white spot moving well ahead of us—that was Ed." He smiled with satisfaction. "We formed a nice community and had good relationships all around, like brothers and sisters—even

though those don't always work!" he laughed.

"That night, I would have liked to enter the minds and dreams of everyone there," Araceli wrote later. "The Sherpas, giving thanks to the divinities that protected them…Ed thinking of Paula; probably that night they both would sleep…David and Robert reviewing each of the decisions and steps they made, and finally feeling happy, congratulating each other…Jamling, traveling the path of his childhood memories, warm from the summit dream that the gods had granted him… and Sumiyo, navigating the sadness of returning home without achieving her dream, but with her strong soul planning her next summit.…"

The team was grateful for Sumiyo's support, and she was happy that they had returned safely. "Now, I look forward to getting away from this small tent and narrow sleeping bag, and to having a hot bath and fresh bedsheets. That is my wish!"

All of them shared excruciating memories of their friends, so recently alive. "This experience might now provide me with a signpost that says: maybe you'll be smart enough to stay away from the mountain for the rest of your life," David remarked, though his eye retained its customary twinkle.

THE HIGHEST WEATHER STATION

The next morning the South Col was clear. Araceli slept soundly, and Ed had to shake her to awaken her. The team managed to fight their exhaustion and hypoxia to break camp, and then assemble the GPS device and weather station.

Moving in slow motion, they hauled the instruments over to a fairly remote site. They

plugged the six sensors into the recording system and used camera tripods to support the anemometer, which indicates wind speed and direction. A separate tripod held the temperature sensors and telemetry antenna. The team deployed the solar panels and then anchored all the parts with rocks.

Then they set up the GPS device, pointed it properly, activated it, and took measurements. The data from the GPS would be used to determine changes in the position and elevation of the South Col since a year earlier when measurements had last been taken. From this data, Roger Bilham would later calculate that the South Col is rising at 4 millimeters per year and moving toward India at 18 millimeters per year.

Establishing "a great moment in Everest history," David Breashears immortalizes the summit scene with his IMAX camera—and 90 seconds of film.

Weather stations sent up in balloons can ascend to higher altitudes, but the station the team fixed, which was tested at temperatures of minus 55°C and in wind speeds of 150 miles per hour, will provide new data on the unusual monsoon-alpine-desert setting of Mount Everest. "It will tell also us about the weather high on the mountain and, with proper evaluation, will make the mountain somewhat safer for climbers," Roger explained. The station also contains a solar radiometer and several temperature probes; all record continuously. The radiometer, which measures incoming radiation, is of interest because of the unit's location closer to the stratosphere than those based at sea level.

The telemetry module can be called up with a

laptop computer and modem, and data can be downloaded. It may eventually be possible to pick up the South Col weather on the Internet.

Although a weather station could be placed on the summit, it is difficult to secure anything to the very top of Mount Everest because the ice that builds up in the winter is ablated and blown away in the summer. "Also, mountaineers hope they are achieving something at the edge of human limitations when they reach the summit," Roger added. "It could be disappointing to find a weather station there."

RETURN TO BASE CAMP
Jamling wasn't able to help with the weather station. When he returned to the South Col from the summit, he had napped briefly, then awoke to realize that he was snowblind. Snowblindness, caused by too much ultra-violet light from the sun—usually exaggerated by reflection from the snow—results in conjunctivitis-like irritation of the cornea. During the climb, Jamling's oxygen mask had fogged his snow goggles, and for much of the climb he hadn't used them.

"That was the most frightening moment of the climb for me—how could I go down? I tried eyedrops, and Sumiyo helped, but I had to break camp and stuff my pack with my eyes closed."

By morning, Jamling could open his eyes for a few seconds, but they quickly teared up. Climbing sirdar Lhakpa Dorje helped him toward Camp II. "On the Lhotse Face, I'd steal a glance for anything dangerous above me, then look down at the route and plunge ahead with my eyes closed. I prayed and thought of my father." He improved at Camp II. Jamling

mentioned the pain only in passing, though snowblindness is acutely painful.

"Returning to Base Camp was a celebration for me," Araceli said. "Some drank beer, but I'd been dreaming about Coca-Cola. I ate so much chocolate that I was too full to eat dinner. Alone in my tent, I thought about all we had done, and I cried, happy that we were safe."

The day the team left Base Camp, May 29, the Sherpas and climbers gathered at the lhap-so and again stoked the juniper incense fires. Along with the team members, they gave thanks to the gods for granting them safe passage, and prayed that they be granted an opportunity to return. They then lowered the *thar-chok* prayer flag pole while a former monk read a prayer. Base Camp had become uncharacteristically quiet.

But another tragedy was about to occur. That day, after the very late hour of 5 p.m., Bruce Herrod, a British climber with the South African team, radioed from the summit. He was never heard from again. Herrod's was the 11th Everest death of the season.

Brad Ohlund was outraged. "How could the South Africans not have learned from what happened just a few days earlier? I'm angry that they allowed a climber to summit so late, and let another person die." The surviving members of the South African team passed through Base Camp on their way down. There were a few nods and hellos, but those still at Base Camp found it difficult to congratulate the South African team members who had returned.

A HERO'S WELCOME
"Araceli! Araceli!" chanted an enthusiastic crowd as Araceli Segarra, the first Spanish

woman to reach the top of Mount Everest, arrived at El Prat airport in Barcelona on June 9. Besieged with attention, she was received by the President of Catalonia, Jordi Pujol.

"I was delivered fan letters addressed simply "Araceli, Alpinist of Everest, Spain." Looking as though she had stepped from a health club, she joked, "I got more rest at Base Camp, where I could sleep without the phone ringing all the time! I lost my voice telling and retelling the story of the expedition. Although some reporters wanted to hear only of the deaths, it was a relief that people in Spain weren't as aware of the May 10 tragedy as in the U.S. When something safe and successful happens, nobody cares, but I tried to convey the good sides to our story."

A PARADE IN DARJILING

A third of the way around the world from Spain, Jamling also received a hero's welcome. "In Kathmandu," he said, "Geshé Rimpoche and Chatral Rimpoche gave me a rare collection of sacred objects to keep on my altar, which will continue to bless me and my family." From there, Jamling left for India, and several hundred relatives and friends greeted him at the border between Nepal and West Bengal. They draped katas around his neck and fed him tea and biscuits. Signs were plastered across the front of several cars, welcoming him home.

The winding streets of Darjiling, about an hour away, were graced with banners. Jamling

Araceli lolls at Camp IV the morning after her ascent. Like most summiters, she moves as if in slow-motion, her mind and body feeling the ill effects of dehydration, hunger, and the low oxygen of high altitude.

FOLLOWING PAGES:
Congratulatory hugs greet the conquering heroes on their return to Base Camp.

was paraded through the hillside town to his family home, where a crowd of Sherpas bearing katas greeted him at the threshold. They also presented him with welcoming *chema*—rice and tsampa held in a divided, ceremonial tray. He took pinches of each and flicked them into the air, as is done at wedding receptions. Beer and chang flowed. "They made me finish an entire bottle of beer before I could enter my own house." A warm welcome evolved quickly into a great party.

"The Darjiling Sherpas were proud of me and proud that I'd advanced my father's name, but mostly I had made them proud of themselves. They were glad that I had been successful on Everest and in America, yet had decided to return to them, to my roots."

Jamling went straight to the altar in the chapel room. His father's thangka scroll painting of the goddess Miyolangsangma rests prominently on the altar, and he did three prostrations, a way of concluding the prayer for good luck and safe travels that he had recited when he departed the house. "I believe that the success and overall safety of our expedition can be partly attributed to the prayers, pujas, lighting of butter lamps, and audiences with lamas," he said thoughtfully. "I feel more devout than I did before the climb. Chomolungma has lit a light within me."

He admitted that his high-altitude mountaineering goals were now largely satisfied. He had climbed the mountain for himself, and to

pay homage to his father, and would now remain at home with his wife and child in Darjiling, running the trekking and climbing business started by his father.

Following his Everest expedition, Jamling joined 350,000 other devout Buddhists in Salugara, West Bengal, for a Kalachakra ("Wheel of Life") initiation presided over by His Holiness the Dalai Lama, who personally blessed him, Soyang, and his daughter.

Jamling was in Darjiling during the fall of 1996 when he heard that Lobsang Jangbu, Scott Fischer's climbing sirdar, had died in an avalanche that swept three climbers from the Lhotse Face. "Lobsang always sought Geshé Rimpoche's blessing before he climbed. But Rimpoche passed away in July, shortly after our spring climb, so Lobsang climbed on Everest in the fall season without Rimpoche's

benediction. He was killed on that climb."

In Kathmandu, Elizabeth Hawley, the renowned Everest statistician, informed Robert that his climb—when compared with his ascent 18 years earlier—marked the longest interval between any individual's two Everest summits.

For a period of time while the team was on the mountain, Robert's family and office staff in his Austrian hometown of Graz were in a panic. The headlines of an Austrian paper had just reported "Styrian Climber Lost on Everest." The article referred to an unnamed mountaineer lost in a tent at 27,000 feet on Everest. An Austrian from Robert's home province of Styria had died on the north side of the mountain, but the climber's identity was not confirmed for almost two days.

When he departed Austria in March, Robert had told few people that he was heading for Everest. He was concerned that press accounts

might raise expectations about his success and create unwanted pressure. He returned to Austria with no fanfare and enjoyed a week of private time before the press found him and coverage of his climb appeared in the media.

Göran Kropp had also accomplished something of a first—he had completed what was certainly the most self-sufficient combined approach and climb of the mountain. David remarked on the conviction, purity, and innocence of the Swede's approach. On his first solo attempt, Kropp turned around when he encountered high winds and deep snow near the South Summit, agonizingly close to the top. His irrepressible enthusiasm was unaffected, and he cheerfully recognized that he would have to accept some assistance and a climbing partner for his two subsequent summit pushes. Kropp said that he intended to bicycle home by a different route, through Russia.

LIFE BEGINS AGAIN

Shortly before Beck Weathers was evacuated from Everest on May 13, Dr. David Shlim, Director of Kathmandu's CIWEC Travel Medicine Center, heard reports that Weathers's arms were frozen up to the elbows. "But when Weathers arrived at our clinic, it was remarkable how good his overall condition was," Shlim said. "Here was the guy I'd been hearing about for two days—first as confirmed dead, then as too ill to risk trying to rescue. Twenty-four hours later he walked into my clinic unassisted."

Makalu Gau was in the main examination room downstairs, unable to walk because of his frostbitten feet, so Shlim settled Beck into the exam room upstairs. First, Beck wanted to call

his wife. Shlim dialed the phone for him, and Beck chatted warmly with a very happy and relieved Peach Weathers.

"Most impressive was Beck's charm, concern for others, and lack of concern for the media," Shlim said. "Initially, he refused interviews, stressing that the real story was the guys who walked him off the South Col and Lhotse Face—Todd, Peter, Ed, Robert, and David."

But a month after Beck returned to Dallas, it became evident that no function would return to his hands. Surgeons there performed state-of-the-art microvascular surgery to save as much of his arms as possible, and he was given skin grafts. Nevertheless, his right arm was dead to his watchband—the one he couldn't gnaw off his arm—necessitating that the arm be amputated in mid-forearm. The fingers of his left hand were dead to the knuckles, including the thumb. Surgeons cut the web of skin between the stump of his thumb and the rest of his hand, giving him limited apposition. He can lay the thumb sideways against his hand and perform tasks such as picking up a piece of paper.

Beck's nose was also destroyed by frostbite. Using cartilage from his ears, some skin, and a piece of rib, doctors rebuilt his nose upside down on his forehead. Once the blood supply was established, they rotated it into position.

In place of his right hand, Beck wears a myoelectric prosthetic device that reads electrical currents from the muscles of his forearm to open and close a claw, allowing him to pour a cup of coffee and pick it up. "When I first wanted to make this trip," he joked, "I looked at the outrageous cost and said, 'You know, this thing is

SHOULD EVEREST BE GUIDED?

The highly publicized deaths of May 1996 have focused unprecedented attention on the limitations and responsibilities of guided climbing.

Many believe that Everest shouldn't be guided for the same reason that the Swiss, with a century of guiding experience, don't generally lead people up the North Face of the Eiger: it's simply too dangerous, for both guides and clients.

In 1985, David Breashears accompanied Dick Bass, 55, to the summit of Everest. Since then, several Everest clients have cited Bass's success as their inspiration. Breashears, however, points out that Bass was not average; he carried his own loads and climbed without ropes where even some experienced climbers have difficulty. "Dick is very independent-minded—he's a powerhouse, a force of nature," Breashears says respectfully. "But for others," he adds, "reaching the top is like stepping up to the plate in the ninth inning of the World Series and hitting the ball out of the park—without having trained, season after season, to acquire the skills needed to get there in the first place."

"The ability to pay the $65,000 fee doesn't make you more experienced," Robert Schauer concurs, expressing a legitimate apprehension about novice clients. But clients come in every variety, from unsea-soned trophy-seekers to experienced Himalayan mountaineers looking for an outfitter and a permit. It should also be remembered that three of the five climbers who died on the south side of the mountain, on or shortly after May 10, were experienced guides. "These deaths did not result from the actions of clients," says client Lou Kasischke, who turned around just below the South Summit.

Kasischke points out that if everything is going right, Everest is very climbable. But if anything goes wrong—and mistakes can be made easily and quickly by even the best—people will die. It is virtually impossible for guides to save anyone above the South Col, a place where every climber must be self-reliant. After Camp IV, the time for "coaching" from the guides is over.

And guides are not paid to take risks. They are paid for their judgment of what risks are worth taking. Indeed, in the mountains, a climber's judgment and attitude are critical. At extreme altitudes, it can be difficult to rationally assess one's physical condition, changes in weather, the effects of delays and other factors. But three of Rob Hall's clients, acting on their own judgment, avoided a life-threatening situation when they decided to turn around.

"Sometimes, climbers will put every bit of their physical and psychological energy into reaching the summit, leaving it to the guides, and others, to get them down," Pete Athans says.

Guide Jim Williams feels that it is time to take a serious look at the responsibility that guides assume when they take money to perform a service. Conceivably, the guiding community could develop some protocols. Tom Hornbein and others feel that, rather than regulation, climber education should be encouraged. Enforcement of regulations, he notes, would be extremely impractical.

Ed Viesturs confirms that he will tighten his own rules for the people he leads. "When guiding, you need to be tough on the clients. I constantly assess their conditions and attitudes, and make certain everyone knows the rules. I won't be flexible on turn-around times, so they have to begin by climbing from one camp to another within minimum times. Summit day is many times harder than any other day on the mountain, so climbers who are having trouble on the earlier days don't belong on the summit ridge." Ed stresses that the fact that people are paying money has never affected the decisions he makes while climbing.

Pete Athans points out that the clients pay him to make his best decisions, but that sometimes means denying them the summit. His associate Todd Burleson concurs, "I feel it's better to have everyone come home safe, but disappointed, than to have one person die." This kind of decision making, in fact, is what most clients are looking for.

"Climbing is no longer only a sport, it has become a business, and for some their only income," concludes Charles Houston. "The more ventured, the more gained." He suggests that this gamble might explain the risks that some guides and clients—possibly for different reasons, and against high odds of weather and terrain—are willing to take on a dangerous mountain such as Everest.

Breashears predicts that within two or three years, the tragedy will be forgotten. His assessment that people will continue to take the same risks, and will inevitably make the same mistakes has been proven right: On a single day in the spring of 1997, five climbers—all on non-guided expeditions—died on the mountain: most of them were climbing late in the day. "I consider myself as a cautious mountaineer," he says. "I don't like to take unnecessary risks, and because of that I might live longer. You can't forsake yourself, or others, for the summit of a mountain."

going to cost me an arm and a leg.' Well, as you can see, I bargained them down."

"Once you've been dead, everything looks pretty good for a while," Weathers said softly. "But it doesn't get much better than coming home to your family. The price that I paid is one thing, but the price my family paid is another. I was able to come back and tell them everything I hadn't been able to before, about how much they mean to me."

"Beck has an incredible will," David said in amazement. "He knew he would lose most of his hands, but he just decided that he wanted to live. He's our one miracle, the one great inspirational story out of all the tragedy—along with

Multiple operations and months of rehab therapy could not save Beck's hands, but did gain him a new nose and improved use of his afflicted limbs.

the selfless heroes among the climbers, Sherpas, Base Camp staff and the helicopter pilot who gave their all to the rescue."

A HEROIC PILOT

As head of his Royal Nepal Army squadron of helicopter pilots, Lt. Col. Madan K. C. felt that if anyone should try a rescue at Camp I, he should. Although most of his flying missions in Nepal are rescues, K. C. said that few survivors have thanked the pilots, and none have written letters. Peach Weathers was an exception, and the letters she wrote poignantly thanking K. C. for giving her husband back to her have helped forge a lasting bond between their families.

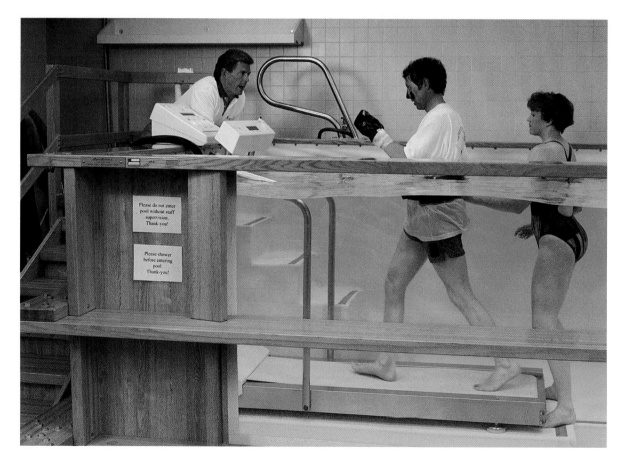

In early 1997, K. C. was invited to the annual meeting of the Helicopter Association International, in southern California. Beck and Peach Weathers were also invited, and when they arrived at the airport in Anaheim, a helicopter identical to the one that rescued Beck was waiting to shuttle him from the airport to the convention center. At the banquet, he saw Madan K. C. for the first time since he had stepped from his helicopter, and they were both near tears. Madan was awarded the Robert E. Trimble Memorial Award, for distinguished service, and he received a standing ovation.

"Madan wasn't acting out of a desire for glory," Beck said. "Until the helicopter set down in Kathmandu, neither of us had any idea that anyone would be the least bit interested in the story."

In April 1997, another function was held at the Smithsonian Institution's National Air and Space Museum, and Beck was selected to present Madan the Aviation Week and Space Technology Laureate Award. In Kathmandu, K. C. was also awarded Nepal's highest honor, the "Star of Nepal" medal, presented by His Majesty the King. "It was not really bravery," the charismatic K. C. said with humility. "It was in the line of duty."

"Through hard work, good judgment and a fair amount of luck, we prevailed," Brad Ohlund reflected. "But not without a stiff reminder of the delicate place man occupies in nature. As for my own involvement, though I never went above Base Camp, I felt privileged, challenged, and humbled."

Filmmaker Greg MacGillivray was upbeat. "We are all proud of the team, and of their determination, courage, skill, and good judgment. Not only did they reach the summit, they had the patience to do it safely."

"Life in the mountains draws out the character of those that journey there," Pheriche Aid Post doctor Jim Litch said. "Maybe this is one of the many reasons why we climb—to see ourselves at the core, not packaged and contained as we are when living within the constraints of technology and consumerism."

"Mountains are hazardous and there will be deaths," Dr. Tom Hornbein concluded, "and as the crowding increases, it's like setting up more pins at the end of the bowling alley: There are more to knock down."

Beck Weathers is sorry that he went to Everest, and he quietly admits, as if ashamed of his former ignorance, that what he was searching for was with him all along: home, health, family, and friends. But like others who have touched Everest's icy skin, he may have come away with something found nowhere else. "One lure of Everest is that the decisions you make there are real, and difficult, and you have to live with them," he reflected. His eyes gleamed almost as if the spirit of Chomolungma were drawing him back. "No doubt, there will continue to be more good men and women willing to gamble and lose for the privilege of being in one of the world's most rare, beautiful places."

FOLLOWING PAGES:

Cast in the glow of a simultaneous sunset and moonrise, this Himalayan skyline boasts (from left) Everest, Lhotse, Ama Dablam, and Makalu: three of these four peaks rank among the five highest in the world.

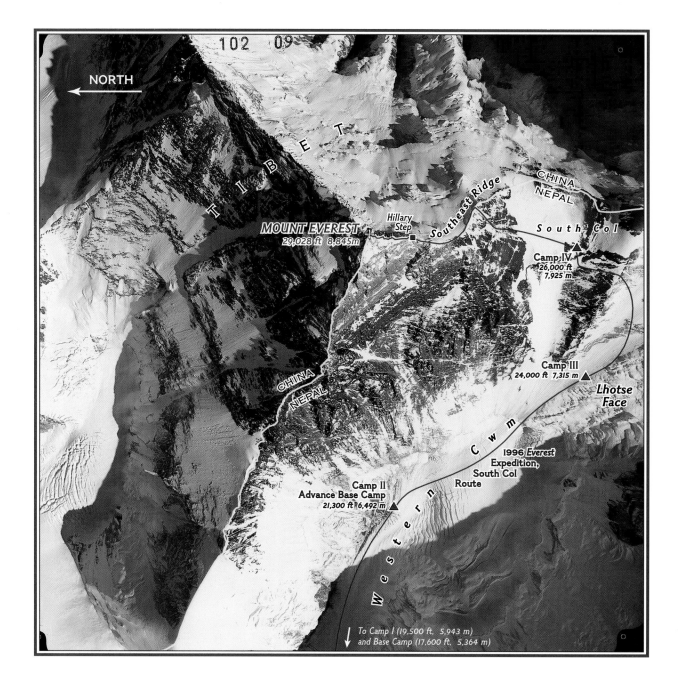

NATURE'S ULTIMATE PYRAMID

*Everest rears three broad walls to the elements: the North, Southwest,
and Kangshung Faces. The route of the* Everest *team closely follows the original,
1953 ascent made by Hillary and Tenzing.*

LIKE RUNNING ON A TREADMILL AND BREATHING THROUGH A STRAW

B Y   D A V I D   B R E A S H E A R S

Of all the icy slopes on Mount Everest, the Lhotse Face is not particularly difficult. A steep incline of hard blue ice, it slants 3,700 feet down the mountainside, requiring little more than competent skills on a fixed rope. I was astonished to hear that someone had just died there. ¶ It was late afternoon May 9, 1996, during my tenth expedition on Everest. The radio call came from Jangbu Sherpa, the head of my camera-carrying team, who had come across an injured Taiwanese climber at Camp III at 24,000 feet. The climber, who had fallen into a crevasse, hadn't complained of serious pains. Yet as Jangbu and two other Sherpas helped him down the Lhotse Face, he collapsed. ¶ "We think he's dead," Jangbu radioed. ¶ I instructed Jangbu to feel the climber's neck for a pulse, to take off his glacier glasses and hold them

243

right under his nose to see if there was any breath.

"No, he's dead."

Superstitious and unwilling to move the corpse, the Sherpas resumed their descent, leaving the body tethered to the fixed rope 1,700 feet above me. I decided to go bring him down. Ed Viesturs and Robert Schauer went with me. Ed, an American climber, was making his fourth ascent of Everest, his third without bottled oxygen. Robert, an Austrian cinematographer, was on the mountain for the second time. They were part of a team under my direction making a film about climbing Everest.

It took us two hours to reach the body. Gasping for breath in the thin air, I kicked the points of my crampons into the slope and slumped back in my harness on the fixed rope to rest. My eyes were riveted on the figure dangling above me. I felt drained, overcome with sadness. The climber's eyes were wide open, his mouth agape, and his face ashen. I later learned he was Chen Yu-Nan, a 36-year-old steelworker from the city of Kaohsiung. When I reached to close his eyes, they had a look of bewilderment, as if from the shock of dying so suddenly.

A few days before in the Icefall, that gargantuan jumble of tilted ice blocks on the Khumbu Glacier, I had watched Chen and his companions as they gingerly placed their feet on the rungs of an aluminum ladder across a crevasse. Their awkwardness showed inexperience. Until recently, I had known most of the climbers on Everest, at least by reputation. They were part of the tightly knit community I had grown up in. But now the routes were crowded with amateurs and guided clients, some of whom had plunked down $65,000 to be led up and down the mountain.

So much has changed since 1983, when I was part of a team transmitting the first video images from the summit. Back then our group of five climbers was the only one on the South Col route. Last year 14 expeditions—South African, Nepalese, Norwegian, Spanish, Swedish, and Taiwanese, among others—shared fixed ropes, camps, and a common obsession to reach the top. Base Camp, once a makeshift staging area at 17,600 feet, has been transformed into a bustling village of more than 300 people, packed with kitchen tents, dining tents, satellite dishes, boom boxes, VCRs, offerings of burning juniper, sputtering generators, and hundreds of prayer flags streaming in the wind. Up above at 26,000 feet, the South Col has been turned into the world's highest garbage dump, with more than a thousand empty oxygen bottles littering the snow alongside torn tents, abandoned stoves, and other refuse.

More than 150 people over the years, including many excellent climbers, have perished on Everest, tumbling from cliffs, being swept away by avalanches, or succumbing to exposure, exhaustion, or altitude sickness. Many of the bodies are still up there. In 1985 I collected body parts from two climbers who had died the year before. Their corpses had frozen solid on the mountain and shattered when they fell to the glacier below.

Last year was the most tragic. In the days that followed Chen's collapse, eight more climbers died in a storm, five on the south side and three on the north. Two were my friends Rob Hall and Scott Fischer, talented guides who knew the mountain well.

In a dark and mysterious way, the deadly nature of the place has only strengthened Everest's grip on the world's imagination. Because the dangers are so obvious, Everest has come to symbolize for many people the ultimate in personal ambition and achievement. Thomas Hornbein, who took part in the first American ascent, in 1963, once described climbing Everest as "a great metaphor for human striving, myth, and the world that is a part of all of us." This explains, in part, why otherwise rational people will pay handsomely to tag the top.

Even veteran Himalayan climbers like myself can find ourselves firmly in the mountain's grip. The risk of death is enticing, because it reminds you that you are alive. "The fact that either you or one of your companions may have the possibility of dying," Sir Edmund Hillary once said, ". . . not only doesn't stop you doing it, but it's almost one of the things that keeps you going." But for me, it's also about the cold, the fatigue, and the challenge of good climbing. It's about the way snow crunches on a minus 10°F morning but squeaks on a minus 20°F morning. It's about moving around a corner and seeing the pink granite of neighboring Makalu glowing in the first rays of dawn.

Everest has this immense psychic gravity that pulls you into its orbit. When George Leigh Mallory and the British reconnaissance team of 1921 set out to find a route up the mountain, it was no more than a set of coordinates on a map. The hulking monster they discovered came as a surprise. "Suffice it to say that it has the most steep ridges and appalling precipices that I have ever seen," Mallory wrote to his wife, Ruth. "I can't tell you how it pos-

sesses me." Three years later, climbing from the Tibetan side, Mallory and his partner, Andrew Irvine, disappeared into the clouds near the summit, never to be seen again. The pair were immortalized by the British press, adding a layer of mystery to the Everest myth.

By the time Tenzing Norgay and Hillary became the first climbers to reach the summit, on May 29, 1953, the Everest story had taken on the hoopla of an international race for the "third pole." Ten expeditions had failed, and 13 men had died. The year before, Tenzing and a Swiss climber had been forced to turn back just short of the summit. News of the British team's dramatic success was flashed to London just as the city was poised to celebrate the coronation of Queen Elizabeth II. "The Crowning Glory: Everest Conquered," the *Daily Mail* proclaimed.

Tenzing's son Jamling was part of our team. Despite his father's urgings not to take up climbing as a career, Jamling had grown up to be an experienced mountaineer and expedition organizer in Darjiling, India. But he had never climbed Everest, and he was determined to reach the summit as part of a lifelong dream and as a tribute to his father, who died in 1986.

I remember as a boy taking a book off the shelf in my family's apartment and turning to the famous photograph of Tenzing standing on the summit. Something fused in my 11-year-old brain as I stared at the Sherpa's thick down suit and overboots, and the ice ax and flags he held aloft in exultation and triumph. I was struck above all by the unwieldy oxygen mask obscuring his face. What kind of place was this, I wondered, where a man needed to carry oxygen to survive?

Today I know the mountain as an environ-

A young monk at Tengboche accepts David Breashears' offer to take a wide-angle gander at his monastery's entryway.

ment so extreme there is no room for mistakes. After the May 1996 storm, members of our film team climbed to Camp III to help nearly a dozen survivors and later managed a helicopter rescue near Camp I. Now, back at Base Camp we were emotionally drained. As we attended an informal memorial for the lost climbers, the summit was flying its pennant-like plume of clouds from the jet stream. At night we lay in our tents listening to the wind on top roaring like a 747 on takeoff. When Jamling called his wife, Soyang, she was deeply worried about his going back up. Ed Viesturs' wife, Paula, our Base Camp manager, was also afraid. She had listened in tears to Rob Hall's last radio calls from the summit ridge and knew that Ed would be climbing without bottled oxygen past the spot where Hall had died with bottled oxygen. Yet we all knew as professionals that we had obligations to make the film, and we were confident in our climbing skills. We agreed to go back.

Just before midnight on May 22, a dozen of us set out from the South Col, climbing by our headlamps and the dim light of the stars. Ed broke a trail for us in the knee-deep snow, a herculean effort. Because we were suffering from the physical effects of the thin air at high altitude, we hadn't slept for more than a few hours in the past three days nor had we eaten more than a few crackers. Our bodies were dehydrated. Our skin was turning blue as precious oxygen was diverted to our brains, hearts, and other vital organs. Climbing above 26,000 feet, even with bottled oxygen, is like running on a treadmill and breathing through a straw. Everything tells you to turn around. Everything says: This is cold, this is impossible. Two hours into the climb, we passed Scott

Fischer's body. Later we found Rob Hall. We kept climbing. By 11 the next morning, we reached the top.

"We can't go any farther," Ed radioed happily to Paula.

As the others celebrated, Robert and I set up the camera. Using my bare hands in the frigid air, I threaded the film through the intricate movement, then looked up to see the photograph from my boyhood coming to life. Holding his ax above his head, Jamling was striking the same pose his father had 43 years before. I was humbled to think of the many journeys I had made to this mountain inspired by that scene.

Several days later, at the bottom of the mountain, I rested on the trail in the rhododendron forest above Tengboche Monastery. In years to come, I knew, the lessons of the tragedy on Everest would be all but forgotten. Climbers would take the same risks, make the same mistakes, and some of them would die, as climbers have been doing for more than seven decades. But smelling the earth and the fragrant trees, I realized in the deepest sense my own good fortune. I had survived Everest once more. I wondered if I would be wise enough to stay away.

Six months later I agreed to make a film on Everest about the effect of high altitude on the body. I would climb the mountain again.

――――――――――――――――――――――

In the winter of 1996, Breashears and Brad Ohlund returned to Kathmandu with the IMAX camera to film some additional air-to-air helicopter shots—with Lt. Col. Madan K.C.

"When your pilot speaks Nepali and English, and the pilot of the other, much larger Mi-17 helicopter is most comfortable with Russian," Brad said, "hand signals only go so far."

"While we were in the air," David added, "K.C. got rescue calls. He'd put me down, I'd offload all the camera gear, and then he'd take off for an hour to rescue altitude-sick trekkers and drop them at Lukla."

The winds generated around Everest and Lhotse are frightening; fixed wing pilots can seldom fly in their vicinity. But one pilot took David to 27,000 feet in the high performance, single engine Pilatus Porter—after David had anchored himself to the floor with the climbing gear that the team had used on Everest. "When we crossed the Lhotse-Nuptse ridge, I slid open the door to a deafening rush of minus 40 degree wind," Breashears recounted. "To keep the camera lenses from fogging, the pilot kindly agreed to turn off the heat, which meant we could keep the door open for only three minutes at a time. At one point I needed to adjust the tripod without my mittens, and the skin on my hands instantly turned bone white."

To keep the wing out of the frame while filming, the pilot banked the plane 30 degrees, which tilted the wing up, then in a risky maneuver applied full right rudder and held that angle as the plane crabbed sideways, shuddering from the stress.

The following spring, Breashears and Ed Viesturs returned to Everest with Pete Athans, Jangbu Sherpa and David Carter to again climb the mountain. They summited and returned safely on May 23, 1997—a year to the day after their 1996 climb.

EVEREST RECORDS

COMPILED BY ELIZABETH HAWLEY

Elizabeth Hawley, a Reuters correspondent and long term resident of Kathmandu, is known in Himalayan climbing circles as the single most reliable source of comprehensive statistics on Himalayan climbing. Ms. Hawley also acts as an executive officer of the Himalayan Trust, and although an American citizen, she is an Honorary Consul to Nepal for New Zealand.

Ms. Hawley is presently working with veteran Himalayan trek leader and computer specialist, Richard Salisbury, on a complete history of mountaineering in the Nepalese Himalayas.

These figures are current as of January 20, 1997.

	EXPEDITIONS	SUCCESSFUL EXPEDITIONS
TOTAL:	391	167
FROM NEPAL:	214	120
FROM TIBET:	176	46

(One expedition successfully climbed both sides simultaneously.)

676 people have stood atop Everest, representing 43 nations and including 39 women. 90 climbers have ascended more than once, including 55 Nepalese Sherpas, 2 Indian Sherpas, and 31 non-Sherpa non-Nepalese, one of whom was an Indian woman, Santosh Yadav.

There have been 846 individual ascents. During the 38-year period from May 1953 to the end of 1991, there were 395 ascents of Everest, or slightly under half of the present total of 846 ascents; over half of the total ascents were made during the 5-year period between the ends of 1991 and 1996.

The largest number of summiters on a single day was 40, on May 10, 1993, all via the South Col-Southeast Ridge route. The spring of 1993 brought the largest number of summiters in a single season: 90, 81 of which ascended from the Nepalese side.

131 successful expeditions traveled 'standard' Everest routes: 98 from Nepal via the South Col to the Southeast Ridge and 33 from Tibet via the North Ridge. One of these teams summited from Tibet and descended the Nepalese side, while another team sent climbers up each side's standard route and down the other simultaneously. The total number of climbers who have summited via the South Col route is well over half, at 445, including 32 women.

60 summiters, including 8 Sherpas and 2 women, said they used no artificial oxygen at any time during their climb. The first to do so were Reinhold Messner, from Italy, with his Austrian partner, Peter Habeler, on May 8, 1978.

MISCELLANEOUS RECORDS

The only complete solo ascent was made by Reinhold Messner on August 20, 1980, via the North Col to the north face and the Great Couloir. The largest number of ascents by a single person is 10, accomplished by Ang Rita Sherpa. His first ascent was in May 1983 and his most recent in May of 1996, at age 49, and he claims that all ascents were made without oxygen.

The fastest ascent was made by Hans Kammerlander, an Italian, on the 24th of May, 1996. His summit, via the standard North Col route, took 16 hours and 45 minutes, from 5:00 p.m. on the 23rd to 9:45 a.m. on the 24th. His climb was timed from Base Camp at 6,400 meters.

Ramon Blanco, a Spaniard but a long-time resident of Venezuela, is the oldest summiter: 60 on October 7, 1993. The youngest was

Nepali Shambu Tamang, who says he was 16 on May 5, 1973. The youngest non-Nepalese was Bertrand Roche, who was 17 on October 7, 1990. He was also half of the first father-and-son pair to summit together. The first son of an Everest summiter to also reach the top was Peter Hillary of New Zealand, on May 10, 1990. Andrej and Marija Stremfelj, now Slovenians, were the first married couple to summit together, on October 7, 1990. The first brothers to the top together were Spanish Basques Alberto and Felix Inurrategi, on September 25, 1992. Robert Schauer of Austria and the *Everest* Film Expedition holds the record for the longest interim between summits: 18 years.

On September 26, 1988, the French climber Jean Marc Boivin became the first to paraglide from the summit. He descended almost 8,000 feet and landed at Camp II in the Western Cwm 11 minutes later. In September of 1992, Pierre Tardivel skied down 10,500 feet from the South Summit, in three hours, the highest ski run ever.

DEATHS ON EVEREST
148 deaths have been reported on Everest, from 22 nations, and including 4 women.

Cause and Location of Death:
97 climbers have died on the Nepalese side, including 19 in the Khumbu Icefall.
Avalanches, falling seracs, or Icefall collapse: 57, of which 36 occurred on the Nepalese side.
From falls: 35, of which 26 occurred in Nepal.
From illness, exhaustion, or exposure: 34, of which 19 occurred in Nepal.
During descent from Everest's summit: 31, of whom 3 were women.
From unknown causes, probably falls: 21, of which 16 occurred in Nepal.

Deaths by Nationality
Nepal: 50, of whom 44 were Sherpas. 41 of these deaths occurred on the Nepalese side, 39 of them Sherpas.
India: 9 Indian Sherpas, 2 of which occurred on the Nepalese side, and 14 non-Sherpas.
Japan: 13

U.K.: 9
China, including Tibet: 8
U.S.: 6
Poland: 5
Czechoslovakia: 5

Incidents with the Largest Number of Fatalities
May 10, 1996: Two New Zealanders, two Americans and one Japanese woman died on the south side from exposure and unknown causes.
May 27, 1989: Five Polish climbers died in an avalanche near the Lho La.
September 9, 1974: One Frenchman and five Nepalese Sherpas were killed in an avalanche.
April 5, 1970: Six Nepalese Sherpas with a Japanese expedition died in a collapse in the Khumbu Icefall.
June 7, 1922: Five Indian Sherpas were killed in an avalanche.

SUMMARY OF DEATHS ON MOUNT EVEREST, SPRING 1996
North Col route:
Indo-Tibetan Border Police Everest Expedition, on May 10-11:
Tsewang Smanla, 38, India
Tsewang Paljor, 28, India
Dorje Morup, 47, India
Reinhard Wlasich, 45, Austria, Hungarian Everest Expedition, May 19

South Col route:

Chen Yu-nan, 36, Chinese Taipei Everest Expedition, May 9
Scott Fischer, 40, USA, Leader, Mountain Madness Everest Expedition, May 10-11
Ngawang Sherpa, Nepal, Mountain Madness Everest Expedition,
Rob Hall, 35, New Zealand, International Friendship Everest Expedition, (IFEE), May 10-11
Andrew Harris, 31, New Zealand, IFEE, May 10-11
Yasuko Namba, 47, Japan, IFEE, May 10-11
Douglas Hansen, 42, USA, IFEE, May 10-11
Bruce Herrod, 37, U.K., Johannesberg Sunday Times Expedition, 25 May

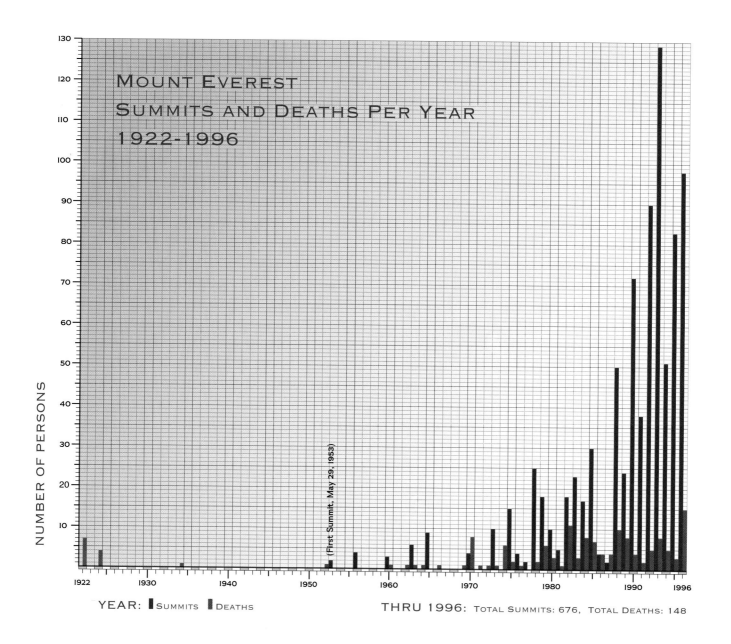

MOUNT EVEREST
SUMMITS AND DEATHS PER YEAR
1922-1996

NUMBER OF PERSONS

(First Summit, May 29, 1953)

YEAR: ■ SUMMITS ■ DEATHS   THRU 1996: TOTAL SUMMITS: 676, TOTAL DEATHS: 148

EVEREST AT A GLANCE

Increasing commercialization of Everest has bred an explosive rise in summiters during the 1980s and 1990s; while deaths also have risen, that curve has been much flatter. Fifteen climbers died on the mountain in 1996— more than in any other year. But when related to the total number who made the top (98), the risk for that year comes across as far less grim than, say, 1970—when only four summited and twice as many perished.

THE 1996 EVEREST FILM EXPEDITION TEAM

**CLIMBERS AND
MOUNTAIN FILM TEAM**

David Breashears,
Newton, Massachusetts, U.S.A.
Expedition Leader and *Everest* film
Co-Director, Co-Producer, and
Cinematographer.

Robert Schauer,
Graz, Austria. Assistant Cameraman.

Ed Viesturs,
Seattle, Washington, U.S.A.
Deputy Leader.

Jamling Tenzing Norgay,
Darjiling, West Bengal, India.
Climbing Leader.

Araceli Segarra,
Catalonia, Spain. Climber.

Sumiyo Tsuzuki,
Tokyo, Japan. Climber.

Base Camp Personnel
Brad Ohlund,
U.S.A. Photographic and Technical
Consultant.

Paula Viesturs,
U.S.A. Base Camp Manager.

Elizabeth Cohen,
U.S.A. Expedition Production Manager.

Audrey Salkeld,
England. Journalist.

Sherpas
Wongchu Sherpa,
Sidar, Chyangba.

Lhakpa Dorje Sherpa,
Climbing Sidar, Chitregaun.

Climbing Sherpas
Jangbu Sherpa, Chyangba.
Muktu Lhakpa Sherpa, Dingjing.
Thilen Sherpa, Dingjing.
Dorje Sherpa, Hosing.
Durga Tamang, Deusa Bogal.
Karsang Namgyal Sherpa, Thame.
Rinji Sherpa, Chyanga.
Ngima Tamang, Gautala.
Ang Pasang Sherpa, Thame.
Ngawang Yonden Sherpa, Pangboche.
Gombu Chhiri Sherpa, Chitre.
Lhakpa Gyalzen Sherpa, Ghat.
Kame Sherpa, Bakam.
Chhuldim Sherpa, Namche Bazar.
Nima Dorje Tamang, Piringding.
Lhakpa Gyalje Sherpa, Chyangba.
Pasang Phutar Sherpa, Gautala.

Non-Climbing Sherpas
Chyangba Tamang, Singati Chhap.
Lhakpa Sherpa, Kerung.
Ngima Sherpa, Chyangba.
Rinji Tamang, Jantar Khani.
Phuri Sherpa, Tingla.

**MACGILLIVRAY FREEMAN FILMS,
LAGUNA BEACH, CALIFORNIA**

Production Team
Greg MacGillivray,
Co-Director and Producer.

Steve Judson,
Co-Writer, Co-Director, Producer, and
Editor.

Alec Lorimore,
Producer.

Kathy Burke Almon,
Production Manager.

Debbie Fogel,
Production Controller.

Book Production
Linda Marcopulos,
Project Manager.

Matthew Muller,
Image (15/70) Reproduction Supervisor.

Chris Blum,
Imagery Researcher.

Film Production/Distribution Team:
Janna Emmel **Myles Connolly**
Teresa Ferreira **Robert Walker**
Bill Bennett **Alice Casbara**

Advisers to the 1996 Everest
Film Expedition
Cynthia Beall,
S. Idell Pyle Professor of Anthropology
and Professor of Anatomy, Case Western
Reserve University.

Roger Bilham,
Professor of Geology, University of
Colorado, Boulder.

James F. Fisher,
Professor of Anthropology and Director
of Asian Studies Carleton College.

Kip Hodges,
Professor of Geology and Dean of
Undergraduate Curriculum,
Massachusetts Institute of Technology.

Charles S. Houston, M.D.

Peter Molnar,
Senior Research Associate,
Massachusetts Institute of Technology.

Audrey Salkeld,
Historian, Author.

Lhakpa Norbu Sherpa,
Former Chief Warden,
Sagarmatha National Park.

Bradford Washburn,
Honorary Director,
Museum of Science, Boston.

Broughton Coburn

IMAX SCREEN SIZE

The 15/70 (IMAX) image is ten times larger than a conventional 35 mm frame and three times bigger than a standard 70 mm frame. The images here represent a comparison of these three film formats when projected onto a theater screen.

70 MM Screen

35 MM Screen

IMAX Screen

**25 m
80 ft**

INDEX

Boldface indicates illustrations

EVEREST
Mountain Without Mercy

BY BROUGHTON COBURN

PUBLISHED BY THE NATIONAL GEOGRAPHIC SOCIETY

Reg Murphy, President and Chief Executive Officer

Gilbert M. Grosvenor, Chairman of the Board

Nina D. Hoffman, Senior Vice President

PREPARED BY THE BOOK DIVISION

William R. Gray, Vice President and Director

Charles Kogod, Assistant Director

Barbara A. Payne, Editorial Director

STAFF FOR THIS BOOK

Kevin Mulroy, Project Editor

Charles Kogod, Illustrations Editor

Michael J. Walsh, Art Director

Anne E. Withers, Researcher

Margery G. Dunn, Consulting Editor

Tom Melham, Picture Legend Writer

Carl Mehler, Senior Map Editor

Joseph F. Ochlak, Map Researcher

Louis J. Spirito, Map Production

Tibor G. Tóth, Map Relief

Richard S. Wain, Production Project Manager

Jennifer L. Burke, Illustrations Assistant

Kevin G. Craig, Editorial Assistant

Peggy J. Candore, Staff Assistant

Anne Marie Houppert, Indexer

MANUFACTURING AND QUALITY MANAGEMENT

George V. White, Director

John T. Dunn, Associate Director

Polly P. Tompkins, Executive Assistant

Composition for this book by the National Geographic Society Book Division. Printed and bound by R.R. Donnelley & Sons, Willard, Ohio. Color Separations by PrintNet/DPI, San Francisco, Calif. Paper by Consolidated/Alling & Cory, Willow Grove, Pa. Dust jacket printed by Inland Press, Menomonee Falls, Wis.

Agleam in sunset, Kusum Kanggru towers over silhouetted trees just off a major trekking trail near the town of Khumjung, not far from the Syangboche airstrip.

ACKNOWLEDGMENTS
AND SUGGESTED READING

It would not have been possible to write this book without the generous cooperation of the climbers, film team, Base Camp staff and academic advisers of the 1996 *Everest* Film Expedition.

In particular, I am indebted to Greg MacGillivray, Alec Lorimore, Teresa Ferreira, Kathy Almon, Myles Connolly, and Matthew Muller of MacGillivray Freeman Films for their help, vision, patience and generous support. I'd like to thank Linda Marcopulos especially for her unending enthusiasm, encouragement and attention to detail—concerning matters legal to grammatical— and to Steve Judson for his careful and skilled attention to the manuscript.

I am grateful to David Breashears and Arcturus Motion Pictures, and National Geographic Society editors Kevin Mulroy and Charles Kogod for their patience and guidance.

To no lesser degree, I appreciate the unswerving help of Liesl Clark of NOVA, Audrey Salkeld, Howie Masters of ABC-TV, Ken Kamler, M.D., Lou Kasischke, Terry Krundick and the staff of Teton County Library, Kevin Craig of National Geographic, Bob Rice of Bob Rice's Weather Window, and my wife Didi Thunder; they deserve special thanks.

Many others offered factual information, critical comments and inspiration, including but not limited to Ian Alsop, Stan Armington, Pete Athans, Myra Badia, Christian Beckwith, Ellen Bernstein, and Encyclopedia Brittanica, Inc., Brent Bishop, Todd Burleson, Brian Carson, Lisa Choegyal, Kate Churchill, Jeanette Connolly, Kanak Mani Dixit, Jenny Dubin, Janna Emmel, Peter Hackett, M.D., Elizabeth Hawley, Thomas Hornbein, M.D., Thomas H. Jukes, Richard J. Kohn, Kevin Kowalchuk of the Imax Corporation, Paul LaChappelle, Wendy Lama, James Litch, M.D., Rick Mandahl, Dave Mencin, Wangchuk Meston, Hemanta Raj Mishra, Bruce Morrison, Brad Ohlund, Brian Peniston, Tom and Sue Piozet, Gil Roberts, M.D., David Schensted, Jeremy Schmidt, Robert Schoene, M.D., Klev Schoening, Pete Schoening, Ang Rita Sherpa, Mingma Sherpa, Phurba Sonam Sherpa, Wongchu Sherpa and Peak Promotions, David Shlim, M.D., Erica Stone, Susy Struble, Norbu Tenzing, Barbara Thunder, Peach Weathers, Seaborn "Beck" Weathers, M.D., Brian Weirum, Jim Williams, and Jed Williamson. I apologize for missing others who also contributed selflessly to this broad-ranging effort.

ABOUT THE AUTHOR

BROUGHTON COBURN *graduated from Harvard College in 1973, and has lived in Nepal and Tibet for 17 of the last 24 years, including three years in Khumbu. He is currently at work on a children's book based on his previous book,* Aama in America: A Pilgrimage of the Heart, *and is writing a book of historical adventure fiction set in the Himalaya. He presently lives in Wilson, Wyoming with his wife, Didi, and their daughter, Phoebe.*

SUGGESTED READING

Vincanne Adams, **Tigers of the Snow and Other Virtual Sherpas;** Jeremy Bernstein, **In the Himalayas: Journeys Through Nepal, Tibet, and Bhutan;** Hugh Downs, **Rhythms of a Himalayan Village;** James F. Fisher, **Sherpas: Reflections on Change in Himalayan Nepal;** Peter Gillman, editor, **Everest: The Best Writing and Pictures From Seventy Years of Human Endeavor;** Carol Inskipp and Tim Inskipp, **A Guide to the Birds of Nepal;** Peter Hackett, **Mountain Sickness: Prevention, Recognition, and Treatment;** Edmund Hillary, **High Adventure;** Tom Holzel and Audrey Salkeld, **First on Everest: The Mystery of Mallory and Irvine;** Hornbein, Thomas, **Everest, The West Ridge;** Hoefer, Hans J., **Nepal;** Charles S. Houston, **Going Higher: The Story of Man and Altitude;** John Hunt, **The Conquest of Everest;** Margaret Jefferies, **Mount Everest National Park;** Patricia Roberts and Thomas Kelly, **Kathmandu: City at the Edge of the World;** Galen Rowell, **Many People Come, Looking, Looking;** Robert Thurman, **Essential Tibetan Buddhism;** James Ramsey Ullman, **Americans on Everest: The Official Account of the Ascent Led by Norman G. Dyhrenfurth;** Walt Unsworth, **Everest: A Mountaineering History** and **Everest: The Ultimate Book of the Ultimate Mountain;** Stephen Venables, **Everest: Kangshung Face;** Michael Ward et. al., **High Altitude Medicine and Physiology;** John West, **Everest: The Testing Place**.

In addition, articles on Himalayan peoples, culture, geology, natural history, and mountaineering can be found in **National Geographic** and other journals such as **Natural History, Himal, American Alpine Club Journal, Mountain,** and **New Scientist**.